# International Conference on Building Materials, Products and Technologies (ICBMPT 2020)

AA007143

IOP Conference Series: Materials Science and Engineering
Volume 1039

Telc, Czech Republic
2 - 4 June 2020

**Editors:**

**Martin Bohac**
**Martina Drdlova**
**Dana Kubatova**
**Theodor Stanek**

ISBN: 978-1-7138-3329-1
ISSN: 1757-8981

**Printed from e-media with permission by:**

Curran Associates, Inc.
57 Morehouse Lane
Red Hook, NY 12571

proceedings.com

**Some format issues inherent in the e-media version may also appear in this print version.**



Institute of Physics
Dirac House, Temple Back
Bristol BS1 6BE UK

Phone: 44 1 17 929 7481
Fax: 44 1 17 920 0979

techtracking@iop.org

**Additional copies of this publication are available from:**

Curran Associates, Inc.
57 Morehouse Lane
Red Hook, NY 12571 USA
Phone: 845-758-0400
Fax: 845-758-2633
Email: curran@proceedings.com
Web: www.proceedings.com

# TABLE OF CONTENTS

## TECHNOLOGY AND TEST METHODS

IOP Conf. Series: Materials Science and Engineering 1039 (2021) 011001 doi:10.1088/1757-899X/1039/1/011001

# Preface

All the world as well as our conference faced to the pandemic in 2020. Therefore, the 23rd International Conference on Building Materials, Products and Technologies (ICBMPT 2020) had to be postponed from the usual date in June to the new date 22nd to 24th September 2020.
The participants appreciated our choice to organize the conference again in the beautiful town of Telč in the Czech Republic, which is popular with many tourists, because it belongs to the cultural heritage under protection of UNECSCO.
Despite the difficult situation, the organizers from the Research Institute for Building Materials collected 24 interesting papers among research workplaces in the Czech Republic and abroad. The conference brings not only interesting presentations, but also it creates a platform for scientific discussions and proposals of new mutual projects.
The main topics of ICBMPT 2020 were: Inorganic binders, Composite materials, Building materials based on secondary raw materials and Technology and test methods in building industry. Also, several keynote speakers were attracted to contribute to our conference. Besides the conference program, there was a possibility to take part in the specialized workshop Healthy buildings with many interesting topics, among others asbestos treatment and biodeteriogens in building environment.
The participants appreciate the personal contact and prefer it to virtual form of participation, therefore, the organizers express their hope that also the next conference will take place in Telč from 1st to 3rd June 2021.
We would like to thank all the participants for their valuable contribution to the conference. We would like to express our gratitude to the reviewers for their time, valuable recommendations and constructive evaluation, in order to guarantee high quality of all the papers. We are looking forward to the next conference in better times!

Scientific committee:

- prof. Ing. Jaroslav Buchar, DrSc., Institute of Thermomechanics, Czech Academy of Sciences
- prof. Ing. Zbynek Keršner, CSc., Faculty of Civil Engineering, Institute of Structural Mechanics, Brno University of Technology
- doc. Ing. Stanislav Rolc, CSc,. Military Research Institute, state enterprise
- doc. Ing. František Šoukal, Ph.D., Faculty of Chemistry, Institute of Materials Science, Brno University of Technology
- RNDr. Theodor Staněk, Ph.D., Research Institute for Building Materials
- Ing. René Čechmánek, Research Institute for Building Materials

Editors:

- Mgr. Martin Boháč, Ph.D.
- Ing. Martina Drdlová
- Ing. Dana Kubátová, Ph.D.
- RNDr. Theodor Staněk, Ph.D.

IOP Conf. Series: Materials Science and Engineering 1039 (2021) 011002 doi:10.1088/1757-899X/1039/1/011002

# Peer review declaration

All papers published in this volume of *IOP Conference Series: Materials Science and Engineering* have been peer reviewed through processes administered by the Editors. Reviews were conducted by expert referees to the professional and scientific standards expected of a proceedings journal published by IOP Publishing.

- **Type of peer review: Single-blind**

Single-anonymous: authors' identities are known to the reviewers, reviewers' identities are hidden from authors

Double-anonymous: author and reviewer identities are hidden to each other

Triple-blind: author and reviewer identities are hidden to each other, and from the Editor(s)

Open: author and reviewer identities are known to each other

- **Describe criteria used by Reviewers when accepting/declining papers. Was there the opportunity to resubmit articles after revisions?**

    - scientific quality
    - style and organization
    - appropriate level of English
    - good level of figures and tables
    - adequate references and discussion to related work

    There was the opportunity to resubmit an article after minor as well as major revisions.

- **Conference submission management system:**

    Each article undergoes assessment by the given editor and then it is sent to 2 or 3 reviewers. According their reviews and recommendations the paper is sent back to the author for revisions. The final approvement of the paper is done by the editor.

- **Number of submissions received:** 30
- **Number of submissions sent for review:** 26
- **Number of submissions accepted:** 26
- **Acceptance Rate (Number of Submissions Accepted / Number of Submissions Received X 100):** 26/30*100=86,67
- **Average number of reviews per paper:** 2,8

- **Total number of reviewers involved:** 58
- **Any additional info on review process (ie plagiarism check system):**

  Each article is checked by a plagiarism detection software (PlagScan).

- **Contact person for queries:**

  Martin Nejedlík, Research Institute for Building Materials, e-mail: conference@vustah.cz

# The role of SCM's on rheology of sprayed mortar

**M Bohac**[1], **D Kubatova**[1], **M Krejci Kotlanova**[1], **I Khongova**[1], **A Zezulova**[1] **and T Stanek**[1]

[1] Research Institute for Building Materials, Hněvkovského 30/65, Brno 617 00, Czech Republic

Email: bohac@vustah.cz

**Abstract**. The role of various SCM's on rheology of fresh sprayed mortar was studied. The mix design was inspired by recent trends in SCM's with use of ground limestone, calcined clays and natural pozzolanas. New formulations were correlated to commonly used SCM's like blast-furnace slag and fly ash. Sprayed mortar designed mostly for thin profile coatings is demanding material in terms of rheology requiring good adhesion, pumpability, shootability, stability after spraying, low segregation and minimum rebound. Particle packing model was used for sprayed mortar mix design. Time and shear rate dependent rheological parameters were calculated using conventional models to describe the properties of fresh sprayed mortars. A procedure for determining the segregation of the mortar using a rheometer was proposed.

## 1. Introduction

Sprayed concrete can be defined as 'mortar or concrete conveyed through a hose and pneumatically projected at high velocity from a nozzle into place.' The material without coarse aggregate is referred to as a *sprayed mortar*. Cement, aggregate, admixtures and water as the constituents of the mortar are mixed together before being fed into the spraying equipment or pump. The mix is then conveyed under pressure to the nozzle, where compressed air is injected to project the mix into place [1]. Fresh mortar composition should meet the requirements for pumpability and shootability.

Yield stress – $\tau_0$ (Pa) and plastic viscosity $\eta_{pl}$ (Pa.s) have meaning only if they can be used to specify certain limits within which the mortar will meet the specific job requirements for workability. The workability and pumpability requirements can be described in terms of slump and water emitted during a pressure bleed test. Pumped mortar moves into the pipeline as a solid plug, which is surrounded by a lubricant layer within which most of the shearing action takes place [2, 3].

Yield stress provides a good explanation of shotcrete shootablility. A low yield stress mortar or concrete would not be suitable for shotcreting; it would simply slough off the receiving surface. On the other hand, mixtures with high yield value (low workability) could be unsuitable for shotcreting, because of pumping and consolidation difficulties [4].

Pumpability and shootability represents practical performance indicators of the wet-mix shotcrete process, both of which are known to have remarkable effects on the quality and efficiency of the shotcreting operation [5].

The study [6], which dealt with rheological properties of fine mortar mixes, describes both the grading of the constituents and the presence of polymers having a significant effect on the flow resistance and torque viscosity. As a part of the study, a rheological audit has been developed and tests for each stage within this audit have been used to characterize the pumpability and sprayability of each mortar.

The properties of fresh wet-mix shotcrete are directly related to shooting process: build-up thickness, rebound and compaction [1]. Morgan [7], when proposing test measurement for build-up thickness quantification, observed two failure mechanisms which may cause the freshly applied concrete to behave in an unstable manner: adhesion failure and cohesion failure. Adhesion can be defined as the ability of shotcrete to adhere to another surface. Cohesion can be defined as self-adhesion, or the ability of the shotcrete to adhere itself. For our application of sprayed mortar in thin layer of few mm the adhesion is crucial parameter. The rebound, as the part of material do not remain in place after the impact, is more relevant for concrete with fibres and coarse aggregate. Segregation and a change in mix composition during shooting with respect to rebound is discussed in [1, 8].

Silica fume replacement in shotcreting operation substantially improved the compressive and flexural strengths of wet-mix shotcrete. Increasing the silica fume content and decreasing the air content led to reductions in permeability, irrespective of the use of air-entraining agents. However, no consistent trend was observed between the spacing factor and permeability of wet-mix shotcrete [9].

Blast-furnace slag (BFS) and fly ash (FA) are less common additives than silica fume, although their inclusion was on the increase, they will became unavailable in near future. Both these traditional SCM's are generally added up to a maximum replacement level of 70% and 35% respectively. They are included to reduce the heat of hydration within thick sprayed concrete applications as they slow down the hydration process. FA can improve the workability and pumpability of a mix made with harsh aggregates or a large proportion of fibres and also provide additional fines to enhance adhesion and cohesion in situations where excessively high cement contents need to be avoided [1].

In case presence with metakaolin authors [10] claim, that mineral admixtures are highly efficient in controlling rebound in dry-mix shotcrete. Particle size of the admixture is of a greater importance than its shape, and finer the particles, greater is the effectiveness of the admixture in controlling material rebound.

The paper summarizes partial results of research and development of special sprayed mortar intended as a thin protective coating for thin-walled sandwich structures. Results of laboratory testing of particle characterization, packing density and rheology are presented. The article gives new information on the role of old and new SCM's, the impact of recycled aggregate on plasticity of fresh mortar and also proposes new approach to the characterization of mortar segregation. Traditional SCM's BFS and FA were compared to future options such as metakaolin (MK) and spongilite (S) as a calcined clay and natural pozzolana.

## 2. Experimental

### *2.1. Materials and methodology*

Commercial materials were used for preparation of mixtures. The binder constituents for sprayed mortar were CEM I 52.5 N (C) (Českomoravský cement, plant Mokrá), finely ground limestone (L) (Vitošov), ground granulated blast furnace slag SMŠ 400 (BFS) (Cemix, plant Kotouč Štramberk), fly ash (FA) (Přerov), metakaolin (MK) (Mefisto K05, ČLUZ), spongilite (S) (Spešov) and silica fume (SF) (Elkem). PCE superplasticizer (SP) Glenium ACE 446 was used for all mixtures. W/c ratio was proposed on same consistency with respect to various water demand of SCM's. Distilled water was used for laboratory tests. Fine fraction of recycled concrete aggregate (RCA) (Beton Brož) was taken as replacement (4 wt.%) for sand. Fine fraction of crushed RCA below 0.355 mm was used (figure 1). Quartz sand of fraction below 1mm was used as aggregate. The ratio of binder to aggregate was 1:2. Sprayed mortars mix design is given in table 1.

**Table 1.** Mix design of sprayed mortars.

| | binder | | | | | | | aggregate | | | |
|---|---|---|---|---|---|---|---|---|---|---|---|
| | C | L | MK | BFS | FA | S | SF | Sand | RCA | SP | w/c |
| Ternary SM | | | | | | | | | | | |
| SM 1 | 100 | | | | | | | 192 | 8 | 1% (b) | 0.4 |
| SM 2 | 55 | 15 | 30 | | | | | 192 | 8 | 1% (b) | 0.46 |
| SM 3 | 55 | 15 | | 30 | | | | 192 | 8 | 1% (b) | 0.37 |
| SM 4 | 55 | 15 | | | 30 | | | 192 | 8 | 1% (b) | 0.4 |
| SM 5 | 55 | 15 | | | | 30 | | 192 | 8 | 1% (b) | 0.43 |
| Binary SM | | | | | | | | | | | |
| SM 6 | 85 | 15 | | | | | | 192 | 8 | 1% (b) | 0.4 |
| SM 7 | 70 | | 30 | | | | | 192 | 8 | 1% (b) | 0.46 |
| SM 8 | 70 | | | 30 | | | | 192 | 8 | 1% (b) | 0.37 |
| SM 9 | 70 | | | | 30 | | | 192 | 8 | 1% (b) | 0.4 |
| SM 10 | 70 | | | | | 30 | | 192 | 8 | 1% (b) | 0.43 |
| Opt. PSD SM | | | | | | | | | | | |
| SM 11 | 86 | 7 | 1 | | 5 | | 1 | 192 | 8 | 1% (b) | 0.4 |

Ideal particle size distribution (PSD) of dry mixtures was observed by correlation of the cumulative grain size curve with the model for ideal distribution according to modified Andreassen model with $q = 0.38$ as in equation (1). The reason why the modified Andreassen model was chosen is that in the past it was often used to characterize of various types of concrete with continuous granulometry, including fine particles. Distribution modulus $q = 0.38$ with its value close to highest possible packing was proposed with respect to lower flowability of fresh mortars.

PSD of the cement, SCM's, RCA and sand was determined by sieve analyses (above 400 μm) and laser particle size analyzer CILAS 920L (0.3–400 μm) with isopropyl alcohol as a dispersing medium. The samples were sonificated for 60 seconds prior to measurement. The data was loaded into EMMA software and the cumulative curves for the selected mixes were compared to the ideal grain size curve of the model. Furthermore, using the available raw materials, a mix was designed that had the best fit with an ideal grain size curve. Best fit was done by simple comparison of various mix designs for each binary and ternary SM with ideal curve.

The shape of RCA was studied in polarized light on optical microscope Nikon Eclipse LV100ND in reflected light at mag. 500x.

$$\frac{CPFT}{100} = \frac{D^q - D_S^q}{D_L^q - D_S^q} \quad (1)$$

where $CPFT$ = cumulative percent finer than $D$, $D$ = particle size, $D_S$ = smallest particle size, $D_L$ = largest particle size, $q$ = distribution modulus.

Mortars were prepared in laboratory at 25°C. Dry binder mixes were homogenized in kitchen aid mixer with flex edge beater. This procedure gives fast homogenization especially for mixes with metakaolin and silica fume. Mortars were prepared in the mixer by adding water with superplasticizer, binder, sand and aggregates and RCA in the end. Fine RCA is added to improve the plasticity of the fresh mixtures.

Prepared mortars were submitted to rheological tests on rotational rheometer DHR-1 (TA Instruments) using geometry of concentric cylinders (DIN). Tests started after 5 min from addition of water. 100 grams of mortar were used in all tests and weighed into a cup (outer cylinder). Testing on the rheometer was performed at 25°C with conditioning of the cup in the jacket before measurement

(Peltier system). Flow parameters were calculated using Bingham [5] and Herschel-Bulkley model as in equation (2, 3). Rheology models were chosen with respect to studied material and its characteristics [11].

$$\tau = \tau_0 + \eta \cdot \dot{\gamma} \qquad \text{Bingham model} \quad (2)$$

$$\tau = \tau_0 + a \cdot \dot{\gamma}^b \qquad \text{Herschel-Bulkley model} \quad (3)$$

where $\tau$ = shear stress (Pa), $\tau_0$ = yield stress (Pa), $\eta$ = viscosity (Pa.s), $\dot{\gamma}$ = shear rate ($s^{-1}$), $a$ = consistency ($Pa.s^n$) , $b$ = rate index.

Diameter and length of inner cylinder was 28 mm and 42 mm, the cup diameter was 44 mm. Samples were pre-sheared with 100 $s^{-1}$ for 10 seconds, logarithmic sweep regimen with 5 points/decade, 5 sec of steady state sensing and 5 sec of averaging time was used. Yield stress - $\tau_y$ (Pa), plastic viscosity $\nu_{pl}$ (Pa.s) and thixotropy/rheopexy (Pa/s) were determined from hysteresis loops in 1 $s^{-1}$ – 150 $s^{-1}$ – 1 $s^{-1}$ region of shear rate. Static and dynamic yield stress were calculated from flow curves (figure 2).

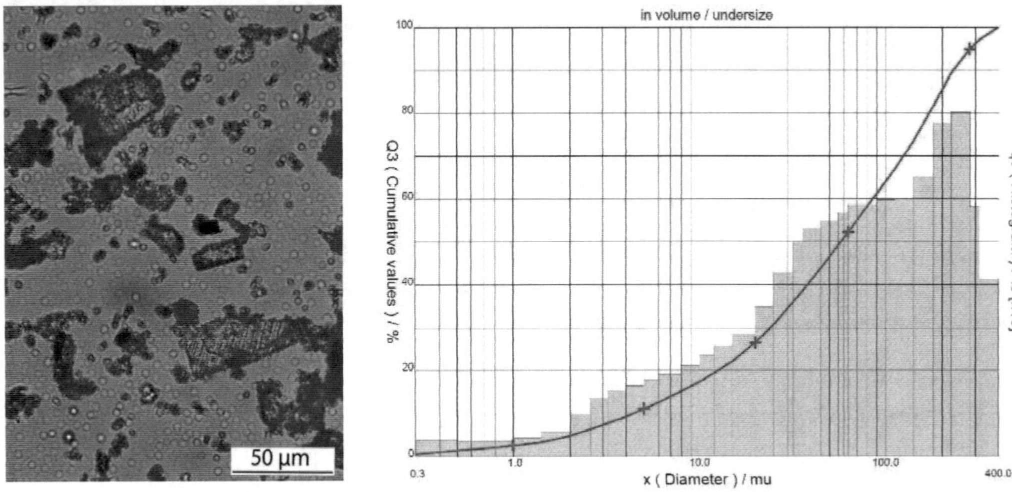


**Figure 1.** Fine fraction of RCA, optical microscopy left, PSD right.

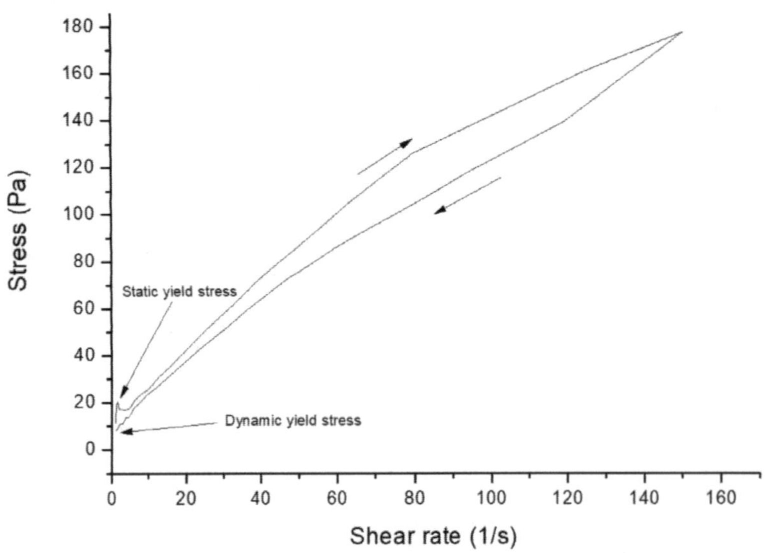


**Figure 2.** Static $\tau_{st}$ and dynamic yield stress $\tau_{dyn}$ (Pa).

## 3. Results and Discussion

### 3.1. Packing density

The grading of mixes was correlated with ideal curve (modified Andreassen model) with distribution modulus $q = 0.38$. Based on model fitting, ideal mix (SM11) was proposed. Ternary and binary mixes age given in figure 3 and 4.

Compared to ideal curve, the reference mix SM 1 show less particles below 10 µm. L, as a component of all ternary binders, slightly improves PSD in the order of microns. MK enriches the particle size range between 5 and 30 µm. The addition of BFS slightly enriches the region below 1 µm. The grading of used FA is similar to PC with slightly more particles below 50 µm. S has higher content of particles between 1 and 50 µm. SF compensates for the lack of particles below 10 µm.

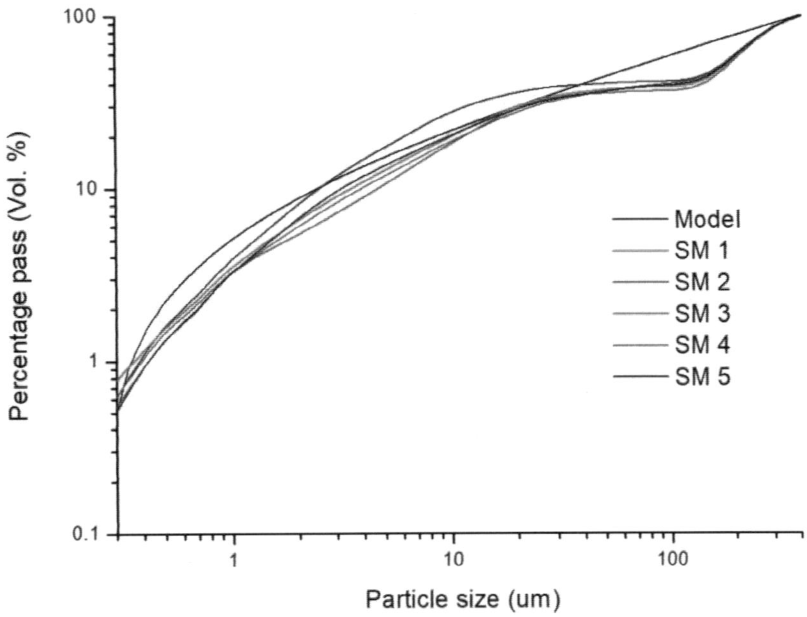


**Figure 3.** Correlation of ternary SM with the model PSD.

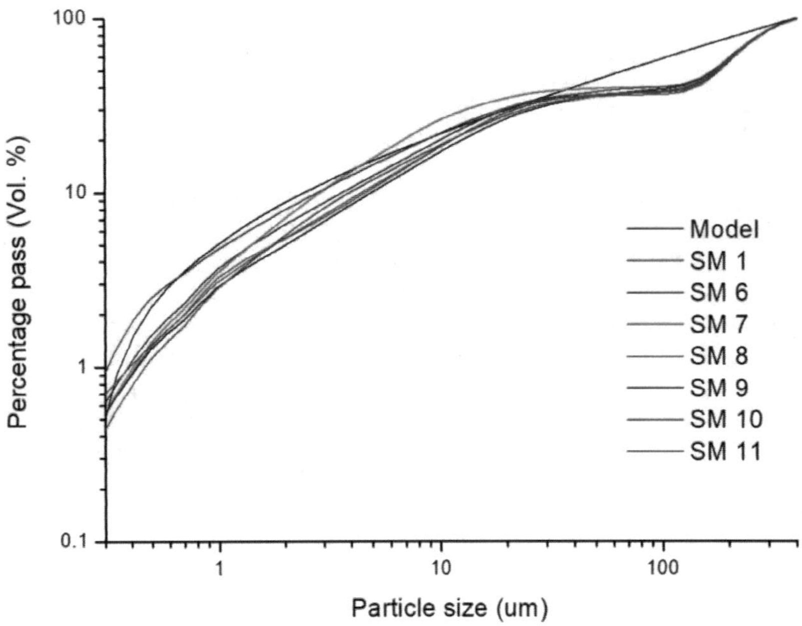


**Figure 4.** Correlation of binary and optimal PSD SM with the model PSD.

The filler effect and improved packing density of used RCA is similar to the effect of granite polishing waste [12], marble dust [13], builder sand or other similar materials. However, RCA added as a replacement for sand in an amount of 4 wt.% to each mix hardly affect the cumulative curve. RCA particles are rather angular than rounded and are not expected to have a positive effect on packing at higher additions. The replacement level of RCA for aggregate was optimized in previous studies [14].

### *3.2. Rheological parameters of sprayed mortars*

The $\tau$ determines how the fresh mixture will hold its shape after spraying and how it will be deformed during the next layering. The $\eta_{pl}$ describes the flow resistance of the fresh mixture. Shear-thinning and shear-thickening as shear-dependent rheological properties are closely related to pumpability and sprayability, mortar segregation and possible clogging of the line, the ability of the sprayed layer to retain its shape, and thus the amount of rebound during spraying [11].

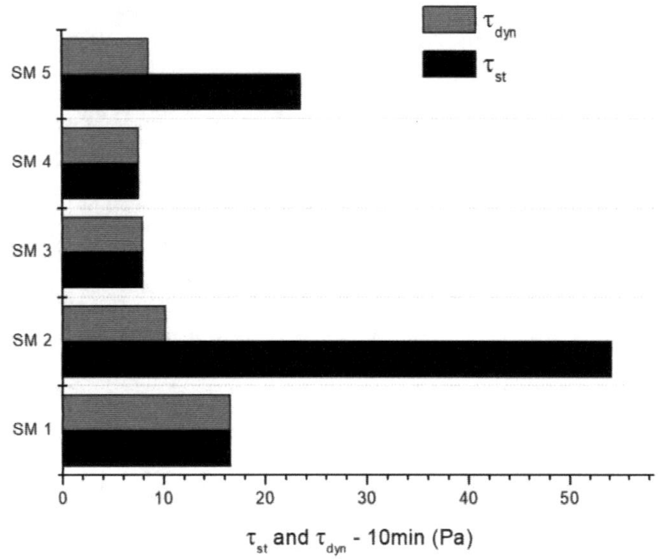


**Figure 5.** $\tau_{st}$ and $\tau_{dyn}$ (Pa) of ternary SM, 10 min after water addition.

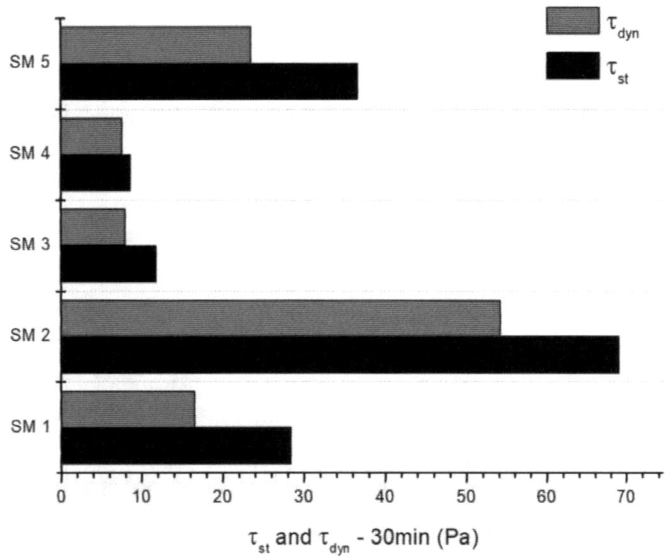


**Figure 6.** $\tau_{st}$ and $\tau_{dyn}$ (Pa) of ternary SM, 30 min after water addition.

Time-dependent rheological properties of thixotropy and rheopexy are important for post-spray properties. Mixtures with relatively high $\tau_{st}$ values and low $\eta_{pl}$ values are suitable for sprayed mortar. The rheological parameters of ternary mixes (figures 5–8) complemented by binary systems were summarized in a rheograph [15] (figure 9) to corelate the effect of individual SCM's by vectorized-rheograph approach.

Comparing reference OPC based SM with binary SM with L, results of rheological parameters show opposite trend than in previous study [16]. One of the reasons is that limestone used in our studied SM was ground separately. This could lead to larger inter-particle spacing compared to systems with interground L. L, which is a component of all ternary binders, reduces $\tau$ and $\eta_{pl}$, when combined with FA. Adverse effect on pumpability related to increased viscosity [17] was not monitored. The spongilite mixture significantly reduces the $\tau$. The fly ash mixture significantly reduces the $\eta_{pl}$, but also the $\tau$. These results are in accordance with previous findings on similar systems [16]. Lower $\tau_{dyn}$ and $\eta_{pl}$ is caused by lower inter-particle friction facilitated by the spherical shape of FA particles. High water demand for MK mixtures and related higher water retention has already been discussed in [18]. Similar effect related to increased water retention was previously described for aerated sprayed mortar with metakaolin [19]. MK increases $\eta_{pl}$ and decreases $\tau_{dyn}$ after 10 min. The values of $\tau_{st}$ are higher than reference mixture. After 30 min, the mixture with metakaolin show significantly higher $\tau_{st}$ and $\tau_{dyn}$. The importance of using high range water reducer admixture to deflocculate fine particle systems with MK and thus affect the water content available to lubricate the flowing system [20] was confirmed in our study. Similar to previous research [17], the BFS does not change the properties of the mixture, but the combination of BFS and L show decrease of $\eta_{pl}$ and $\tau_{dyn}$ (figure 6, SM 3). Based on $\eta_{pl}$ after 10 min and 30 min, the faster setting was observed for the mixture with L and BFS. A possible cause is an increased content of fine particles.

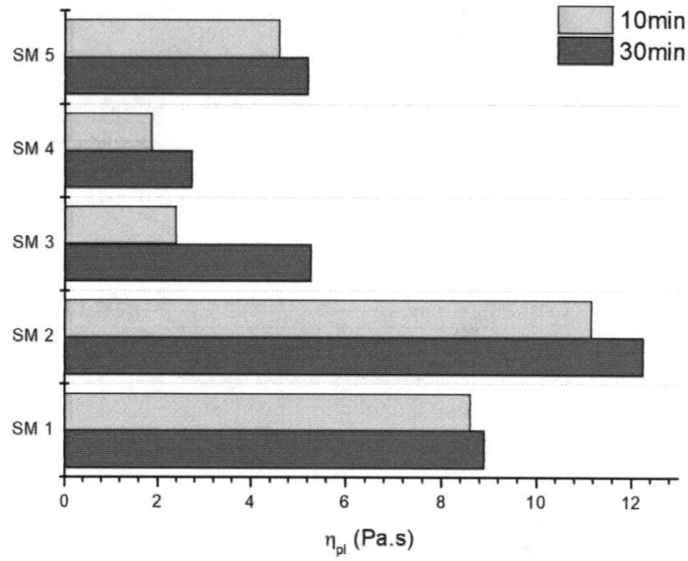


**Figure 7.** $\eta_{pl}$ (Pa.s) of ternary SM, 10 min and 30 min after water addition.

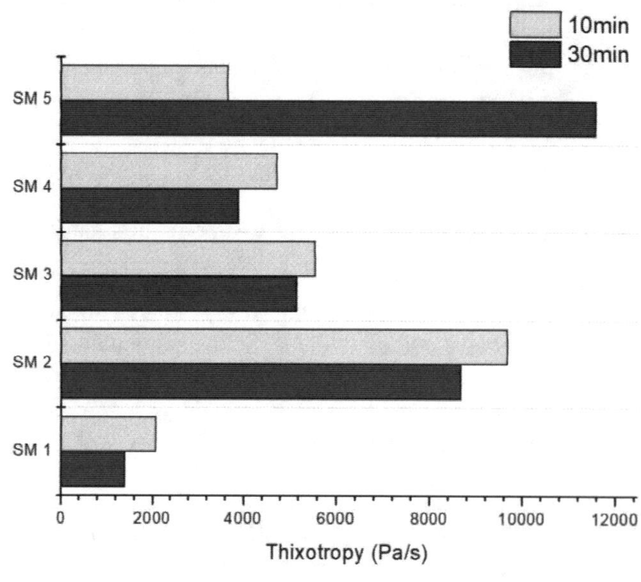


**Figure 8.** Thixotropy (Pa/s) of SM, 10 min and 30 min after water addition.

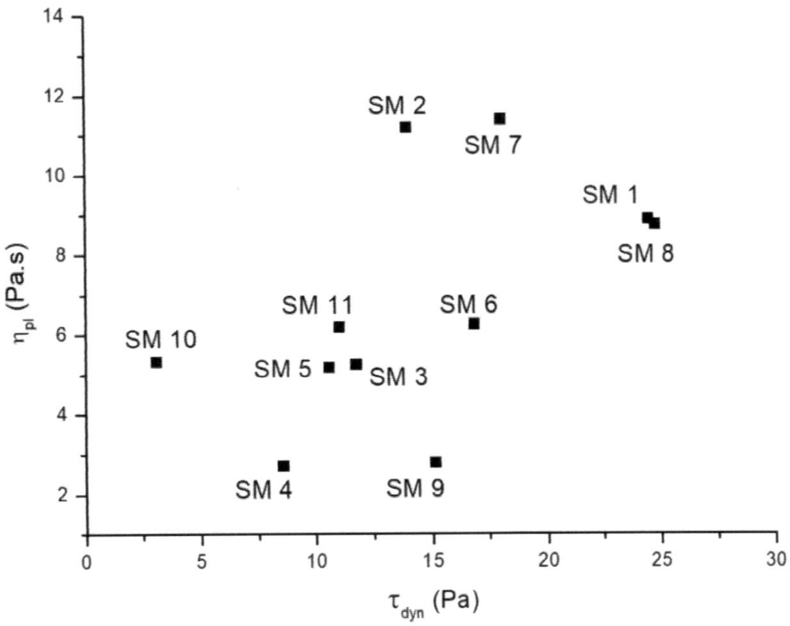


**Figure 9.** Rheograph $\eta_{pl}$ / $\tau_{dyn}$ of spraying mortars with binary and ternary SM (10 min).

Mixtures with MK and S have higher $\tau_{st}$ and thixotropy values. MK mixtures have very plastic behaviour and good adhesion. A mixture with FA and BFS has a suitable low viscosity needed for good pumpability. Similar to 3D printing, RCA addition in sprayed mortars decreases workability but improves the plasticity of fresh mixtures [14].

### *3.3. Segregation test on rheometer*

Segregation resistance is very important feature of a sprayed mortar fresh mixture. Segregation of concrete can be determined by standard [21], but there are no standards for mortar.

Segregation of binder from aggregate in fresh mortar is related to pumping process. Separation of the paste from the aggregates related to segregation can lead to blockage of the line. Coarser particles are continuously forming the plug that eventually causes the blockage [22]. An analogy to sprayed mortar can be found in shotcrete, where lower cohesion related to lower yield stress has almost proportional relationship to build-up thickness [5, 11]. While yield stress can be approximated from the flow curve at low shear rates, segregation can be observed at higher shear rates.

The flow curve should be designed to simulate shear rates during material application. As the velocity increases, the stress in the material to be measured is released at a certain shear rate by expelling the water-containing material, in this case a paste, towards the points from the higher shear rate, i.e. to the edge. It is therefore an analogous process that takes place during segregation when pumping of sprayed mortar. In this case, the use of concentric cylinder geometry is justified, as the test simulates the passage of a mortar through hose with a smooth surface of the inner tube.

On the ascending flow curve, segregation can be observed and quantified as pseudoplasticity. Pseudoplasticity is the flow behaviour of non-Newtonian fluids, characterized by a decrease in the large initial apparent viscosity with increasing shear stress.

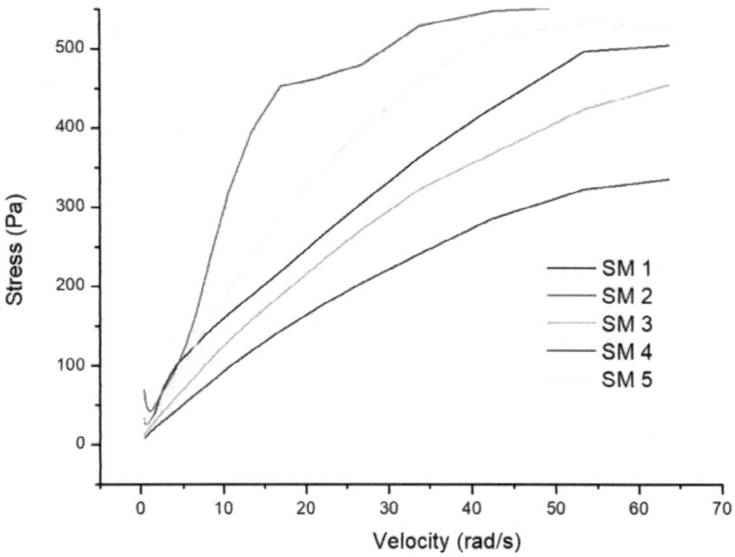


**Figure 10.** Ascending flow curve of sprayed mortars with binary binders.

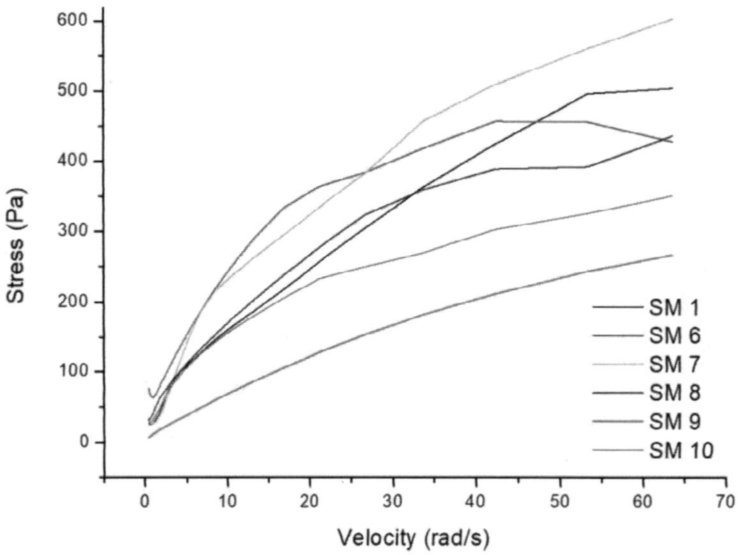


**Figure 11.** Ascending flow curve of sprayed mortars with ternary binders.

Segregation can be quantified from the flow curve in two ways; as a critical shear rate or as an exponent of the Herschley-Bulkley model (rate or flow index). Neither method is sufficiently accurate and one or the other is more suitable for different materials. While some materials experience shear-thinning behaviour gradually from low shear rates, other materials with sudden change in stress at a certain shear rate or we monitor combination of both. To illustrate the segregation, the rate index from Herschel-Bulkley model was used for studied sprayed mortars (figure 10–12). The mortars with a higher rate index have greater segregation resistance and are therefore more suitable for pumpability without risk of blockage problems. Reference SM1 and SM with FA and combination of L and FA or BFS show good segregation resistance. Poor segregation resistance was monitored for L and S and combination of L and MK.

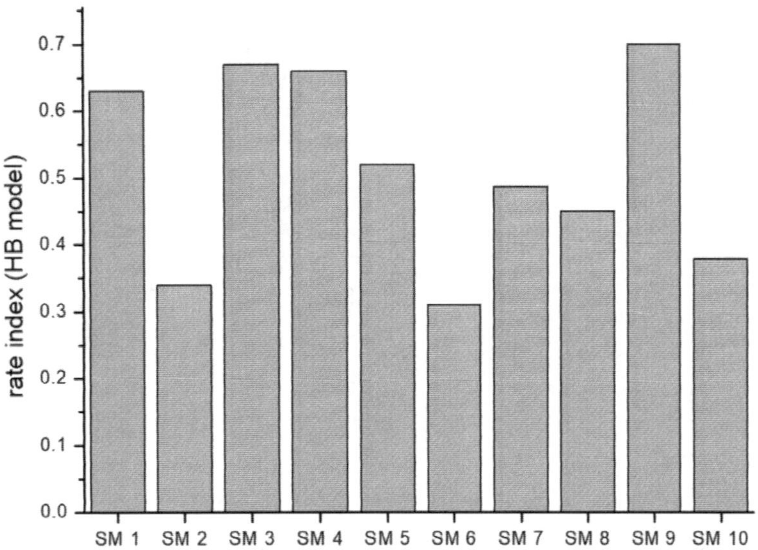


**Figure 12.** Rate index, sprayed mortars with binary and ternary binders.

## 4. Conclusion

Special care has to be taken when designing mixes with SCM's for sprayed mortar. Segregation resistance as the key parameter of fresh sprayed mortar is closely related to proper grading that prevents the blockage of the spraying system. Adhesion, pumpability, shootability and stability after spraying of sprayed mortars can be judged based on rheological parameters.

The lack in particles below 10 μm in reference sprayed mortar can be compensated by SF and BFS. MK introduces particles in the range 5–30 μm. Spongilite has higher content of particles between 1 and 50 μm.

The use of L reduces $\eta_{pl}$ and $\tau_{dyn}$, when combined with FA without adverse effect on pumpability. The FA significantly reduces both $\eta_{pl}$ and $\tau_{dyn}$. Addition of MK increases values of $\tau_{st}$ and decreases workability time. Addition of BFS does not cause considerable changes in properties of the fresh mixture, but the combination with L the $\eta_{pl}$ and $\tau_{dyn}$ decreases. Short workability time was revealed for combination of L and BFS.

Mixtures with MK and S have higher static yield stress and thixotropy values showing plastic behaviour and good adhesion. Addition of combination of FA and BFS leads to mortar with low viscosity needed for good pumpability. Addition of fine fraction of RCA to sprayed mortar mix decreases workability but improves the plasticity of fresh mixture.

Segregation can be quantified as rate index on flow curve. The mortars with a higher rate index show greater segregation resistance and good pumpability. Mixtures with FA show good segregation resistance. Mortars with combination of L and MK can cause blockage problems due to lower segregation resistance.

**Acknowledgements**
Thanks are due to Institutional support for long-term development of research organizations by the Ministry of Industry and Trade of the Czech Republic.

**References**

[1] Austin S A, Robins P J and Goodier C I 2002 Construction and repair with wet-process sprayed concrete and mortar *Shotcrete Magazine* **4**(1) pp 10–12
[2] Tattersall G H and Banfil P F G 1983 The Rheology of Fresh Concrete *Pitman, London*, 365 p.
[3] Chen L, Liu G, Cheng W and Pan G 2016 Pipe flow of pumping wet shotcrete based on lubrication

layer *SpringerPlus* 5:945

[4] Baupré D, Mindess S and Morgan DR 1993 Development of high Performance Shotcrete (theoretical considerations) *Engineering Foundation Conference, Shotcrete for Underground Support VI Niagara-on-the-Lake, Canada, May 2–6 1993* pp 1–8

[5] Yun K-K, Choi P and Yeon J H 2015 Correlating rheological properties to the pumpability and shootability of wet-mix shotcrete mixtures *Construction and Building Materials* (98) pp 884–891

[6] Austin S A, Robins P J and Goodier C I 1999 The rheological performance of wet-process sprayed mortars *Magazine of Concrete Research* (51–5) pp 341–352

[7] Morgan D R 1991 High Early Strength Blended-Cement Wet-Mix Shotcrete *Concrete International* Vol 13 No. 5 pp 35–39

[8] Parker H W 1977 A Practical New Approach to Rebound Losses, in ACSE and ACI SP-54, Shotcrete for Ground Support, Detroit pp 149–187

[9] Choi P, Yeon J H and Yun K-K 2016 Air-void structure, strength, and permeability of wet-mix shotcrete before and after shotcreting operation: The influences of silica fume and air-entraining agent *Cement and Concrete Composites* (70) pp 69–77

[10] Bindiganavile V and Banthia N 2001 Fiber reinforced dry-mix shotcrete with metakaolin *Cement and Concrete Composites* (23) pp 503–514

[11] Liu G, Cheng W, Chen L, Pan G and Liu Z 2020 Rheological properties of fresh concrete and its application on shotcrete *Construction and Building Materials* 243 118180

[12] Chen J, Li B H, Ng P L and Kwan A K H 2020 Adding granite polishing waste as sand replacement to improve packing density, rheology, strength and impermeability of mortar *Powder Technology* **364** pp 404–415

[13] Li L G, Huang Z H, Tan Y P, Kwan A K H and Chen H Y 2019 Recycling of marble dust as paste replacement for improving strength, microstructure and eco-friendliness of mortar *Journal of Cleaner Production* **210** 55/65

[14] Bohac M, Nespor B and Stanek T 2018 The role of various aggregates on rheology of fresh 3D printing concrete *Proceedings of SynerCrete'18 International Conference on Interdisciplinary Approaches for Cement-based Materials and Structural Concrete, 24–26 October 2018, Funchal, Madeira Island, Portugal* pp 843–848

[15] Wallevik O H and Wallevik J E 2011 Rheology as a tool in concrete science: The use of rheographs and workability boxes *Cement and Concrete Research* **41** (2011) pp 1279–1288

[16] Mehdipour I, Kumar A and Khayat K H 2017 Rheology, hydration, and strength evolution of interground limestone cement containing PCE dispersant and high volume supplementary cementitious materials, *Materials & Design* **127** (2017) 54–66.

[17] Choi P, Yun K-K, Yeon JH 2017 Effects of mineral admixtures and steel fiber on rheology, strength, and chloride ion penetration resistance characteristics of wet-mix shotcrete mixtures containing crushed aggregates *Construction and Building Materials* **142** pp 376–384

[18] Melo J P, Aguilar A S and Olivares F H 2014 Rheological properties of aerated cement pastes with fly ash *Construction and Building Materials* **65** pp 566–573

[19] Melo J P, Flores Medina N, Sepulcre Aguilar A and Hernández Olivares F 2017 Rheological and thermal properties of aerated sprayed mortar *Construction and Building Materials* **154** pp 275-283

[20] Paiva H, Velosa A, Cachim P and Ferreira V M Effect of metakaolin dispersion on the fresh and hardened state properties of concrete *Cement and Concrete Research* **42** (2012) pp 607–612

[21] CSN EN 12350-11 (731301) Test for fresh concrete - Part 11: Self-compacting concrete - Test for segregation during sieving

[22] Browne R and Bamforth P B 1977 Tests to Establish Concrete Pumpability *ACI Journal* Vol. 74 No. 5 pp 193–207

IOP Conf. Series: Materials Science and Engineering 1039 (2021) 012002 doi:10.1088/1757-899X/1039/1/012002

# Preparation, manufacture and properties of new β-belite cement

**J Kohout[1], P Koutník[1] and A Soukup[1]**

[1] Unipetrol Centre for Research and Education, a.s., Revolucní 1521/84, Ústí nad Labem, 400 01

Email: jan.kohout@unicre.cz

**Abstract.** The objective of this work is the optimization of the synthesis of β-belite cement. Belite cement (dicalcium silicate) is a promising material due to a lower $CO_2$ emission and lower energy consumption during its production. Cement with a high content of β-belite was prepared from ground limestone, amorphous silica and alkaline silicate solution. The raw materials were mixed, the mixtures were compacted and then burnt in an electric furnace. Different $CaO/SiO_2$ ratios, firing temperatures, raw materials and firing time were tested. The samples were characterized by XRD, SEM and free lime content. Differences between the cements with high content of β-belite prepared on a laboratory scale (tens of grams) and on a larger scale (tens of kilograms) were studied. Belite cement prepared on a larger scale had better properties compared to a laboratory scale. Optimal synthesis of β-belite cement consisted of mixing ground limestone, silica fume and liquid potassium silicate and burning at temperature 1100°C for more than 2 hours.

## 1. Introduction

Belite is dicalcium silicate, $Ca_2SiO_4$. It occurs in multiple forms (α, β and γ). The occurrence of α, β or γ form of belite is determined by the temperature of the firing during its manufacture. Belite in the forms α and β is an important component of Roman cement. A small amount of belite, in the γ form is also found in Portland cement, but the main component of Portland cement is tricalcium silicate, alit, $Ca_3SiO_5$. $CO_2$ emissions in belite production are stoichiometrically a third lower than the $CO_2$ emissions found with the formation of alite, as they arise according to the equations (1) and (2):

$$2\,CaCO_3 + SiO_2 \rightarrow Ca_2SiO_4 + 2\,CO_2 \qquad (1)$$

$$3\,CaCO_3 + SiO_2 \rightarrow Ca_3SiO_5 + 3\,CO_2 \qquad (2)$$

The production of belite cement is also much less energy-intensive than the production of Portland cement, as belite cement is produced at temperatures of 800 to 1100°C, while Portland cement at temperatures above 1400°C. This will result in further reductions in $CO_2$ emissions at a significant fuel savings rate. Production is therefore also more economically advantageous.

Methods of the preparation of belite cement or pure belite in laboratory scale have been studied more intensely in the recent decades. The most widespread process involves the mixing of powdered raw materials, agglomeration and final burning of the mixture [1, 2]. Dispergation of the raw materials in the water or ethanol provides a better homogeneity of the mixture [3, 4]. Another common methods of the preparation of belite cement or pure belite are a hydrothermal method [5, 6] or sol-gel process [7, 8].

In a previous study, a simple procedure was developed to prepare cement based on β-belite [9]. The procedure was based on solid-state reaction, using pure raw materials (powdered limestone, amorphous silica and liquid alkali silicate) produced industrially in large quantities, which ensures the constant quality of the materials. The alkali silicate was used as a source of reactive $SiO_2$ and of alkali ion and for his ability to bind effectively powdered raw materials. The procedure was aimed at preparing β-belite at a fixed stage at high temperatures. Since the preparation of units of grams of belite cement in laboratory conditions, tens of kilograms have been prepared. The new procedure of preparation of β-belite cement is protected by a patent CZ 306912 B6 [10].

The aim of this work was therefore to verify the possibility of producing belite cement on a larger scale and to compare its properties with the properties of cement produced from the same raw materials by analogue process on a scale of several orders smaller.

## 2. Experiment

### *2.1. Raw materials and methods*

*2.1.1. Raw materials.* For the preparation of β-belite in small quantities, the following basic raw materials were used. Silica fume (České lupkové závody, a.s., Czech Republic), limestone (Omyacarb 5VA; Omya) and potassium silicate (Water Glass, a.s., Czech Republic). All raw materials were described and analysed in detail. Bulk materials using XRF, XRD, BET, SEM, particle size distribution, bulk volume mass, humidity and annealing loss were determined. For water glass, the total water content, density, silica content and the aggregate content of alkaline metal oxides were determined by titration. The ICP-OES method determined the oxide content of other elements and the ratio of potassium oxides to sodium.

The chemical composition and physical properties of solid materials and potassium water glass are presented in table 1 below.

**Table 1.** Results of raw material analyses.

| description | limestone | silica fume | potassium silicate |
|---|---|---|---|
| $SiO_2$ [%] | 0.50 | 95.6 | 21.29 |
| $Al_2O_3$ [%] | 0.27 | 0.23 | 0.03 |
| $Fe_2O_3$ [%] | < 0.01 | 0.30 | < 0.01 |
| CaO [%] | 55.1 | 1.34 | < 0.01 |
| MgO [%] | 0.41 | < 0.01 | < 0.01 |
| $K_2O$ [%] | < 0.01 | < 0.01 | 8.18 |
| $Na_2O$ [%] | < 0.01 | < 0.01 | 0.70 |
| $P_2O_5$ [%] | < 0.01 | 0.44 | < 0.01 |
| $ZrO_2$ [%] | < 0.01 | 1.14 | < 0.01 |
| $SO_3$ [%] | < 0.01 | 0.11 | < 0.01 |
| humidity [% r.h.] | 0.07 | 0.55 | not specified |
| loss on ignition [%] | 43.69 | 0.68 | not specified |
| $D_{50}$ [μm] | 6.27 | 3.80 | not specified |
| $D_{90}$ [μm] | 17.67 | 13.61 | not specified |
| BET [$cm^2/g$] | 6.77 | 15.32 | not specified |
| bulk density [$kg/m^3$] | 787 | 303 | not specified |
| total water content [%] | not specified | not specified | 69.02 |
| specific gravity [g/cm3] | 2658 | 2204 | 1.327 |

IOP Conf. Series: Materials Science and Engineering 1039 (2021) 012002 doi:10.1088/1757-899X/1039/1/012002

*2.1.2. Analytical and testing methods.* The chemical compositions of powdered raw materials were determined by X-ray fluorescence (BRUKER S8 Tiger).

A BRUKER D8 Advanced X-Ray diffraction system (XRD) equipped with a BRUKER SSD 160 detector and operating with Cu-Kα radiation at 40 kV and 25 mA was used for analysis of raw materials and prepared dicalcium silicates. XRD scanning was taken at the $2\theta = 0.02$ step over an angular range from 5° to 70° with 1 s counting time. The crystalline phases were identified using the powder diffraction files from database ICDD - International Centre for Diffraction Data PCPDFWIN v.2.2. (2001).

A Mastersizer 2000 laser diffraction particle size analyser (MALVERN Instruments) was used to determine size distribution of powdered raw materials. Agglomerates were disrupted by ultrasound treatment.

A gas sorption analyser Autosorb iQ from Quantachrome was used for the determination of specific surface area by the Brunauer-Emmett-Teller method (BET).

Free lime content in dicalcium silicate samples was determined by a glycerine-alcohol test.

The morphology of belite cement at a laboratory scale was studied by a scanning electron microscope Mira 3 from TESCAN. For SEM analysis of belite cement on a larger scale, the sample was coated with about 5 nm of gold to make it conductive under the microscope JEOL JSM-IT500HR. Representative secondary electron images of the microstructure were taken at 7000x magnifications in high vacuum mode using an accelerating voltage of 10 keV and a working distance of 10.3 mm.

An inductively coupled plasma optical emission spectrometer OPTIMA 8000 (Perkin Elmer) was used to determine the content of micro-elements and K/Na ratio in liquid alkali silicates. Total content of alkali metals (Na, K) and content of $SiO_2$ in alkali silicates were determined by conventional acid-base titration methods; the reason for their application being higher accuracy at higher concentrations compared with other methods.

Specific gravity was determined by the pycnometric method.

### *2.2. Procedure*

*2.2.1. Laboratory preparation procedure.* Raw materials were used in a weight ratio of silica : limestone : water glass 100 : 454 : 175. The $CaO/SiO_2$ molar ratio was 2 mol/mol. Actual preparation consisted of homogenizing bulk components by hand in a beaker. The bulk components were thermal silica and ground limestone. Then potassium water glass was poured into the beaker and the mixture was further mixed by hand, homogenized. Finally, 20 g of the mixture was compressed into tablets with a diameter of 40 mm and with a pressure of approx. 20 MPa for 2 minutes using a hand hydraulic press. The resulting tablets (agglomerates) were calcined in an electric laboratory furnace at 1100°C for 8 hours. Cooling was spontaneous in the furnace. In the end, the calcinates were crushed in a friction bowl and ground in a planetary mill. The above conditions of preparation were selected as optimal on the basis of previous work [1].

*2.2.2. Preparation procedure in larger quantities.* The preparation of larger quantities was based on optimised preparation of belite cement in small quantities. Thus, such raw materials were used and procedures were chosen that allowed β-belite to be produced in the best quality. The same raw materials were used to produce cement in larger quantities as in small quantities. Cement was therefore made of ground limestone, thermal silica and potassium water glass. Raw materials were used in a weight ratio of silica : limestone : water glass 100 : 454 : 175. The $CaO/SiO_2$ molar ratio was 2 mol/mol.

Dry ingredients were homogenized in a mixer type nautamix (Nauta-mixer). This homogenized mixture was sprinkled into a planetary mixer on potassium water glass. The mixture of all raw materials was homogenized in the mixer for 10 minutes. The prepared mixture was compacted by hand into a ball shape with a diameter of approx. 60 mm. These compacts were calcined in an electric furnace at 1100°C for 8 hours. After firing, the calcinates spontaneously cooled in the closed furnace. After cooling,

IOP Conf. Series: Materials Science and Engineering 1039 (2021) 012002 doi:10.1088/1757-899X/1039/1/012002

the balls were crushed with the help of a jaw crusher and further ground in a ball mill. In the end, all cement was homogenized. A total of 25 kg of belite cement was produced.

## 3. Result and discussion

Comparing the resulting properties of belite cements on a scale of tens of grams and on a scale of tens of kilograms was the basic goal of this work. The results of XRD, SEM and free lime concentrations were compared. Physical properties were evaluated only for cement produced in larger quantities. Cement produced on a laboratory scale was too little for physical analysis.

### *3.1. XRD*

From the diffractogram shown in figure 1 below, it is clear that the belite cement contained dicalcium silicate mainly in the form of ß-belite (larnite, ICDD 33-0302). A small amount of another form of dicalcium silicate ($C_2S$, ICCD 31-0297) was also present. No other phases (α-belite, γ-belite, wollastonite, etc.) were identified in the prepared belite cement compared to other studies [4, 5, 11–14]. The resulting diffractograms show that samples of belite cement produced in small and larger quantities have almost the same phase composition.

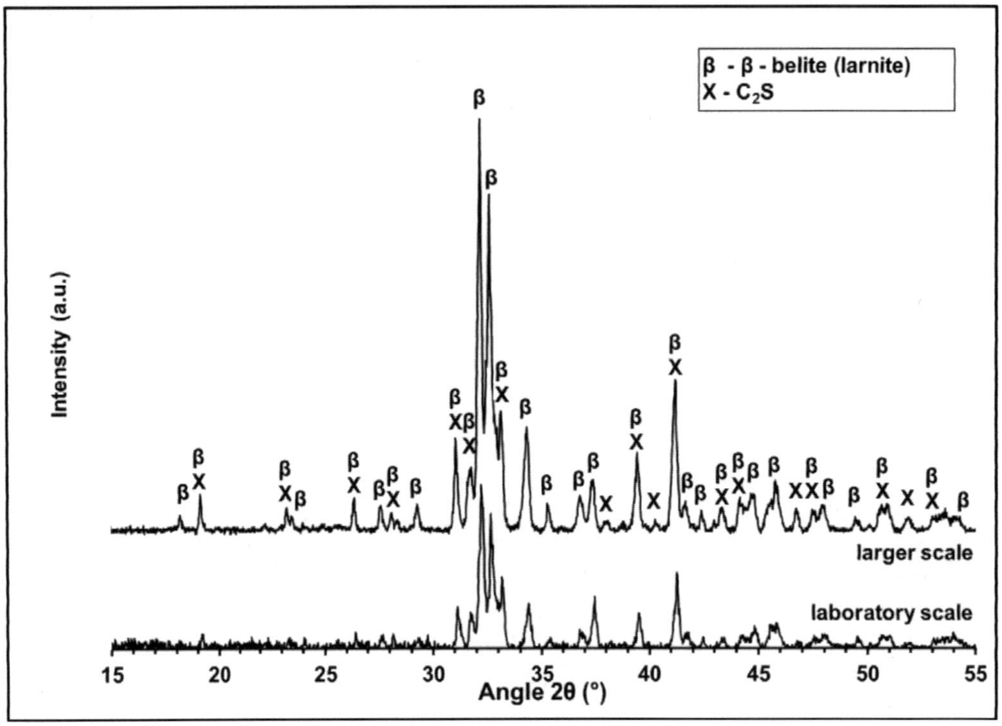


**Figure 1.** Comparison of diffratograms of belite cements produced on a laboratory scale and on a larger scale.

### *3.2. Free lime*

The mass concentrations of free lime are given in table 2. The mass concentration of free lime in the belite cement sample produced in larger quantities is much less than the free lime concentration in the belite cement sample produced in small quantities. This difference was probably caused by different homogenization of dry components (ground limestone and thermal silica) during production. Machine homogenization during production of higher volumes of belite cement was better than manual homogenization in the production of belite cement in small quantities. Similar results of free lime in the belite cement were observed by El-Didamony et al. [12] (approx. 2%) and Mazouzi et al. [13] (approx. 2%).

**Table 2.** Mass concentration of free lime in samples of belite cements.

| β-belite cement | free lime [% wt.] |
|---|---|
| **laboratory scale** | 4.19 |
| **larger scale** | 1.63 |

### *3.3. Physical properties*

The results of physical analyses are presented in table 3. Due to the very small amount of belite cement sample produced on a laboratory scale, physical properties were determined only for belite cement produced in larger quantities. Belite cement produced in larger quantities was very sintered as it had a specific surface of only 3.8 $m^2/g$ and its specific gravity was 3.3 $g/cm^3$. After grinding, the vast majority of particles were below 100 μm in size. Maheswaran et al. [14] had a comparable result of specific gravity (3.16 $g/cm^3$) of belite cement prepared from calcined lime sludge and silica fume. The result of specific surface area of belite cement was lower compared to other study [5]. The difference was probably caused by the presence of potassium ions (in the liquid alkali silicate) that support the sintering.

**Table 3.** Physical properties of belite cement produced in larger quantities.

| description | belite cement |
|---|---|
| **$D_{50}$ [μm]** | 5.20 |
| **$D_{90}$ [μm]** | 70.5 |
| **BET [$m^2/g$]** | 3.8 |
| **bulk density [$kg/m^3$]** | 886 |
| **specific gravity [$g/cm^3$]** | 3.303 |
| **total pore volume [$mm^3/g$]** | 435 |
| **average pore size [nm]** | 607 |

### *3.4. Surface morphology*

The particles of belite cement and their surface were examined in detail using a scanning electron microscope (SEM) and are shown in figure 2 (laboratory scale) and figure 3 (larger quantities). The morphology of the particles of the belite cement sample produced under laboratory conditions and the sample of cement produced in larger quantities were compared. There are no differences between belite cement on a laboratory scale and belite cement on a larger scale. The particles of both cement samples had a smooth sintered surface. The particles were individual without impurities.

**Figure 2.** Micrograph of belite cement on a laboratory scale.

**Figure 3.** Micrograph of belite cement on a larger scale.

## 4. Conclusions

A new process was developed to produce cement with a high content of ß-belite. Production was very simple and consisted of mixing industrial raw materials normally available on the market. The amount of cement produced was gradually increased up to tens of kilograms. The resulting belite cement produced in larger quantities showed better properties than cement produced in laboratory quantities. Analyses showed high ß-belite content and low free lime content. Cement produced in larger quantities contained more than 2.5 times less free lime than cement made on a laboratory scale. The resulting cement was of a very bright color and with a high content of ß-belite, therefore it is a very promising material for the production of building materials, especially for the restoration of monuments.

In the following works, the application of belite cement, i.e. its hydration properties and the properties of mortars based on it, will be verified.

**Acknowledgement**

The author wishes to thank Karol Bayer (Faculty of Restoration, University of Pardubice) for providing SEM pictures using an electron microscope from TESCAN. This publication is a result of the project Development of the UniCRE Centre (LO1606) which has been financially supported by the Ministry of Education, Youth and Sports of the Czech Republic (MEYS) under the National Sustainability Programme I. The result was achieved using the infrastructure of the project Efficient Use of Energy Resources Using Catalytic Processes (LM2015039) which has been financially supported by MEYS within the targeted support of large infrastructures.

## References

[1] Chen Y L, Lin Ch J, Ko M S, Lai Y Ch and Chang J E 2011 Characterization of mortars from belite-rich clinkers produced from inorganic waste *Cement and Concrete Composites* **33** pp 261–266

[2] Staněk T and Sulovský P 2015 Active low-energy belite cement *Cement and Concrete Research* **68** pp 203–210

[3] Kriskova L, Pontikes Y, Zhang F, Cizer Ö, Jones P T, Van Balen K and Blanpain B 2014 Influence of mechanical and chemical activation on the hydraulic properties of gamma dicalcium silicate *Cement and Concrete Research* **55** pp 59–68

[4] Morsli K, De la Torre A G, Cuberos A J M, Zahir M and Aranda M A G 2009 Preparation and characterization of alkali-activated white belite cements *Materiales de construcción* **59** pp 19–29

[5] Kurdowski W, Duszak S and Trybalska B 1997 Belite produced by means of low-temperature synthesis *Cement and Concrete Research* **27** [1] pp 51–62

[6] Garbev K, Beuchle G, Schweike U, Merz D, Dregert O and Stemmermann P 2014 Preparation of a Novel Cementitious Material from Hydrothermally Synthesized C-S-H Phases *Journal of the American Ceramic Society* **97** [7] pp 2298–2307

[7] Dovál M, Palou M and Kovár V 2006 Hydration and microstructure of binder compounds containing $C_2AS$ and $C_2S$ synthesized by sol-gel method *Ceramics-Silikáty* **50** [2] pp 106–114

[8] Stephan D and Wilhelm P 2004 Synthesis of Pure Cementitious Phases by Sol-Gel Process as Precursor *Zeitschrift für anorganische Chemie* **630** [10] pp 1477–1483

[9] Koutník P 2017 Preparation of β-belite using liquid alkali silicates *Materiales de Construcción* **67** 328 e140

[10] Koutník P 2017 Způsob výroby belitického cementu *Patentový spis CZ 306912 B6 26.7.2017* (Ústí nad Labem: Unipetrol Výzkumné vzdělávací centrum a.s.)

[11] Morsli K, De la Torre A G, Stöber S, Cuberos A J M, Zahir M and Aranda M A G 2007 Quantitative Phase Analysis of Laboratory-Active Belite Clinkers by Synchrotron Powder Diffraction *Journal of the American Ceramic Society* **90** [10] pp 3205–3212

[12] El-Didamony H, Khalil Kh A, Ahmed I A and Heikal M 2012 Preparation of β-dicalcium silicate

(β-C2S) and calcium sulfoaluminate (C3A3 CS) phases using non-traditional nano-materials *Construction and Building Materials* **35** pp 77–83

[13] Mazouzi W, Kacimi L, Cyr M and Clastres P 2014 Properties of low temperature belite cements made from aluminosilicate wastes by hydrothermal method *Cement and Concrete Composites* **53** pp 170–177

[14] Maheswaran S, Kalaiselvam S, Saravana Karthikeyan S K S, Kokila C and Palani GS 2016 β-Belite cements (β-dicalcium silicate) obtained from calcined lime sludge and silica fume *Cement and Concrete Composites* **66** pp 57–65

IOP Conf. Series: Materials Science and Engineering 1039 (2021) 012003 doi:10.1088/1757-899X/1039/1/012003

# Raman spectroscopy of interfacial transition zone in concrete doped with limestone powder and metakaolin

**M Krejcí Kotlanova**[1] **and V Masek**[2]

[1] Research Institute of Building Materials, Hnevkovskeho 65, 617 00, Brno, Czech Republic

[2] Palacky University, Institute of Molecular and Translation Medicine, Hnevotinska 5, 779 00, Olomouc, Czech Republic

Email: krejci@vustah.cz

**Abstract**. The interfacial transition zone in concrete with ordinary Portland cement (OPC) and concrete containing OPC and doped with 20% of limestone powder and 30% of metakaolin was studied by Raman spectroscopy - optical profilometry and cluster analysis. Mechanical properties, including strength characteristics and absorbability were tested after 28 days of curing. Concrete with ordinary Portland cement has lower density and absorbability and higher compressive and tensile strength than concrete with limestone powder and metakaolin. It has been found that a large amount of portlandite is formed in the interfacial transition zone in concrete with ordinary Portland cement. In concrete with the addition of metakaolin and limestone powder, the thickness of the interfacial transition zone is reduced, less portlandite is produced, but a thin layer of calcite is formed around quartz grains. This results in lower concrete strength in the area of the interfacial transition zone.

## 1. Introduction

Nowadays, there is the effort to reduce $CO_2$ in production of concrete. So, Portland cement can be partly replaced by supplementary cementitious materials (SCM´s), such as metakaolin, diatomite, silica fumes, limestone powder, pulverized fly ash or ground granulated blast furnace slag [1, 2].

SCM´s have a lot of advantages for cement industry. First is economic aspect of the matter – replacing Portland cement by industrial by-products. Second advantage is environmental aspect – to reduce greenhouse gases associated with Portland cement production. And third, performance advantage – SCM´s can greatly improve the workability and durability of concrete [3].

Clays and limestone are a strategic raw material available worldwide. Recent research [4] focused on the study of limestone calcined clay cement ($LC^3$), which contains 40–50% of clinker, 30–40% of calcined clay, 15–20% of limestone powder and 4–7% of gypsum, has shown that $LC^3$ has comparable development of compressive strength as ordinary Portland cement (OPC), higher resistance to penetration of chlorides, moisture, gas and sulphate attack, higher corrosion resistance and lower porosity. Nevertheless, the highest disadvantage is lower resistance to carbonation [5].

### *1.1. Interfacial transition zone*

The interfacial transition zone (ITZ) is a weak area between cement paste and aggregate particles (figure 1). The thickness of this zone is mostly from 15 to 50 μm in concrete with ordinary Portland cement [6]. The ITZ is more porous than the cement paste and has high w/c. In the ITZ, crystallization of larger portlandite crystals, ettringite and very porous hydration products occur [7, 8]. C-axes of ettringite crystals are often oriented perpendicular to the aggregate surface [9].

Recent research [10] has shown that ITZ is formed when the packing of anhydrous cement grain is disrupted in the area between filler and binder. The ITZ change can therefore occur when we change the particle size distribution of cement materials. The addition of SCM´s to binder can improve the resulting concrete properties and reduce the thickness and porosity of ITZ. Only a few authors studied the ITZ using Raman spectroscopy [e.g. 11, 12, 13], but none of them studied ITZ in concrete with the addition of metakaolin and limestone powder.

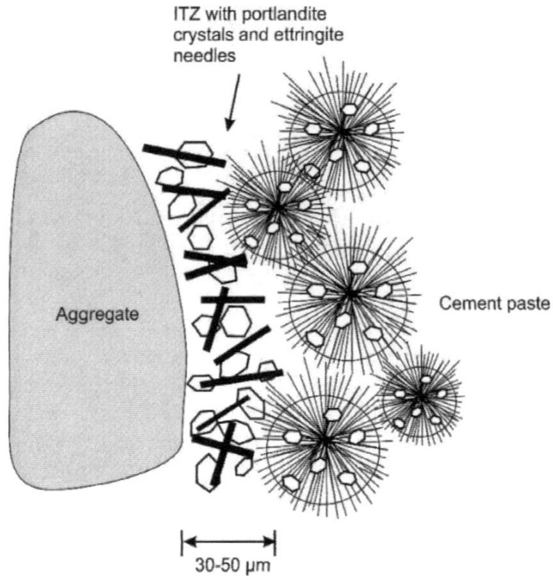


**Figure 1.** Schematic diagram of ITZ in concrete with ordinary Portland cement according to Al Bayati et al. [14], modified.

## 2. Materials and methods

Concrete sample OPC-1 was prepared from ordinary Portland cement 52.5 N. River sand from locality Bzenec was used as fine-grained aggregate and crushed granitoids from the Olbramovice quarry were used as coarse-grained aggregate. Fractions 0/2 and 2/4 were used in the mixture (table 1). The binder for VPC-1 sample was prepared from 50% of ordinary Portland cement, 30% of metakaolin and 20% of limestone powder. The filler was the same as in the case of OPC-1. According to XRD analysis, the used metakaolin Mefisto K05 has the following phase composition: amorphous phase 70.5%, mullite 7.7%, microcline 6.8%, illite/muscovite 4.9%, quartz 4.7%, kaolinite 4.2% and anatase 1.2%. Mixtures were casted in 40 × 40 × 160 mm moulds and cured for 28 days in a bath of distilled water. Then mechanical properties were measured on samples. After that, a small block was cut from both samples and the samples were ground. Subsequently, Raman spectroscopy measurements were done.

Mechanical properties, including strength characteristics and absorbability were tested at Research Institute of Building Materials in Brno, Czech Republic. Raman spectroscopy of selected samples was performed at Institute of Molecular and Translation Medicine at the Palacký University in Olomouc, Czech Republic on a WITec Confocal Raman Imaging Microscope System alpha300 R + with excitation of 532 nm (power incident on the sample - 35 mW, objectives Zeiss EC Epiplan 20x/0.4 and Zeiss EC Epiplan-Neofluar Dic 50x/0.8, scan bitmap (step) = 1 pixel/μm, integration time at each point = 200 ms. Mapping area was 200 × 200 μm for sample OPC-1 and 400 × 400 μm for sample VPC-1. Before the Raman measurement, the selected area was scanned using optical profilometry WITec TrueSurface.

**Table 1.** Mixture proportions for OPC-1 and VPC-1.

| | OPC-1 | VPC-1 |
|---|---|---|
| CEM I (52.5 N) [g] | 450 | 225 |
| Limestone powder - Štramberk [g] | - | 90 |
| Metakaolin Mefisto K05 [g] | - | 135 |
| River sand - Bzenec [g] | 450 | 450 |
| Aggregate 0-2 mm - Olbramovice [g] | 450 | 450 |
| Aggregate 2-4 mm - Olbramovice [g] | 450 | 450 |
| Water [g] | 273 | 355 |

## 3. Results

### *3.1. Mechanical properties*

Mechanical properties, including strength characteristics and absorbability were tested after 28 days of curing. Concrete with ordinary Portland cement has lower density and absorbability and higher compressive and tensile strength than concrete with limestone powder and metakaolin (table 2).

**Table 2.** Mechanical properties of OPC-1 and VPC-1 samples in 28 days.

| | Density (kg/m$^3$) | Tensile strenght (MPa) | Compressive strenght (MPa) | Absorbability (%) |
|---|---|---|---|---|
| OPC-1 | 2056 | 7.3 | 51.0 | 10.3 |
| VPC-1 | 2168 | 5.6 | 27.0 | 15.6 |

### *3.2. Raman spectroscopy*

Raman spectroscopy was performed on both samples. Suitable binder-filler interfaces were selected for study and then a cluster analysis was performed, in which the individual average spectra of the measured area are categorized into clusters with different spectra properties. Due to the imperfections of the sample surface, the optical profilometry method was chosen, when the laser beam copies the real sample surface, not as in conventional Raman analysis, where the beam analyses the plane, but doesn´t solve the roughness of the sample.

*3.2.1. Sample OPC-1.* The boundary between feldspar and cement paste (figure 2a) was selected for study by Raman spectroscopy. Analysis distinguished seven clusters (figure 2b) differing in their average spectra. Real profile of the sample was analysed by optical profilometry and is shown on figure 2c. Cluster analysis delimited the ITZ quite well, three clusters define ITZ and the thickness of this zone is between 5 and 45 μm. Raman spectra for concrete with OPC are shown on figure 3. C-S-H is a major hydration product of Portland cement in concrete and corresponds to the bending mode at 645-654 cm$^{-1}$. Portlandite was found in all the three clusters, in contrast, ettringite wasn´t detected in any of these clusters. A large amount of iron oxides is formed in the ITZ, iron can come from C4AF present in the OPC. It´s not easy to distinguish some phases since two or more phases may have some similar peaks. Raman band at 1000 cm$^{-1}$ in brown spectra (figure 3) probably corresponds to some sulphates or C-S-H.

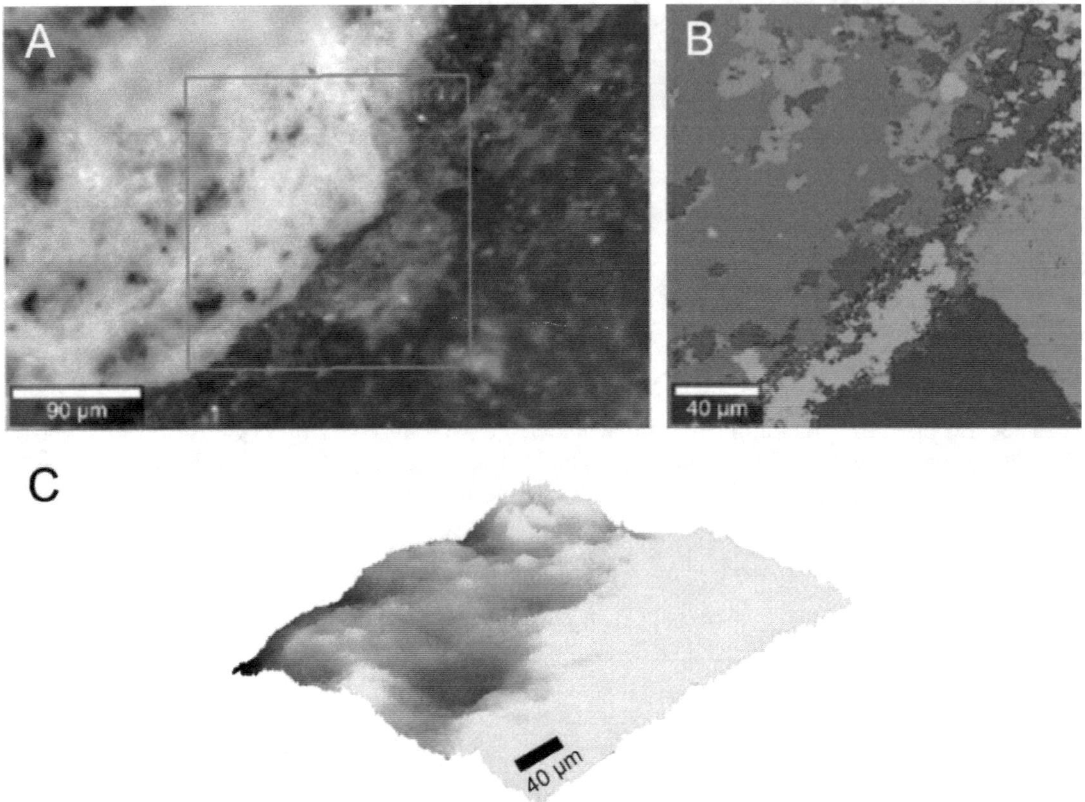


**Figure 2.** Concrete sample OPC-1 **A)** Bright-field image under the confocal optical microscope; **B)** Raman image of the ITZ of the sample; **C)** Sample profile analysed by optical profilometry.

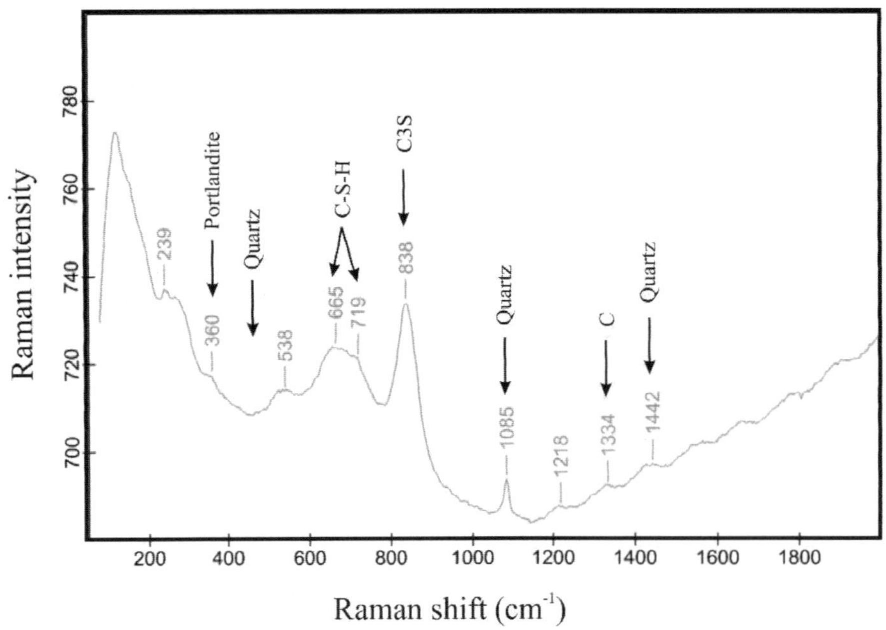


**Figure 3.** Raman spectra for individual clusters in sample OPC-1.

IOP Conf. Series: Materials Science and Engineering 1039 (2021) 012003 doi:10.1088/1757-899X/1039/1/012003

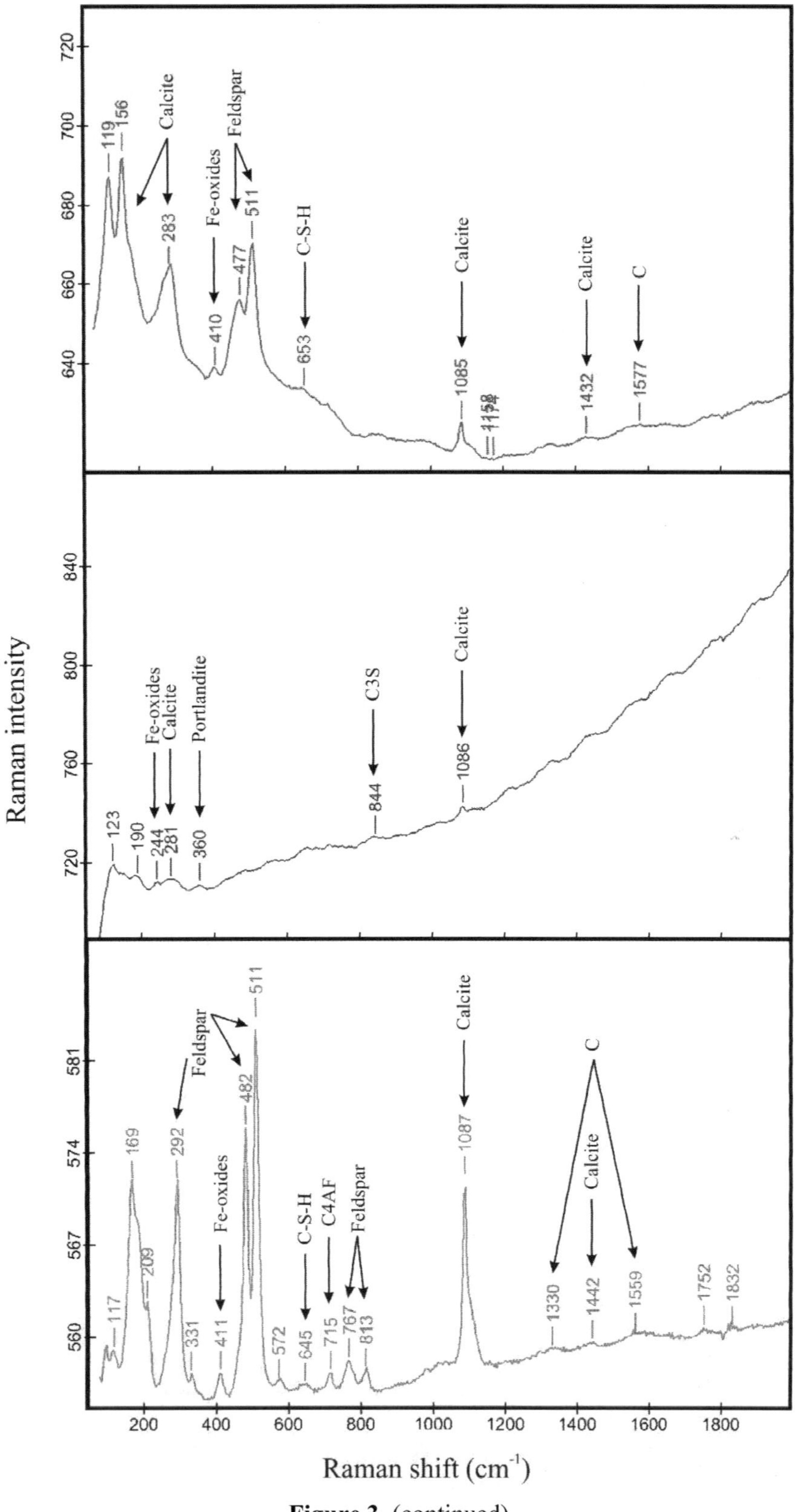


**Figure 3.** (continued).

IOP Conf. Series: Materials Science and Engineering 1039 (2021) 012003 doi:10.1088/1757-899X/1039/1/012003

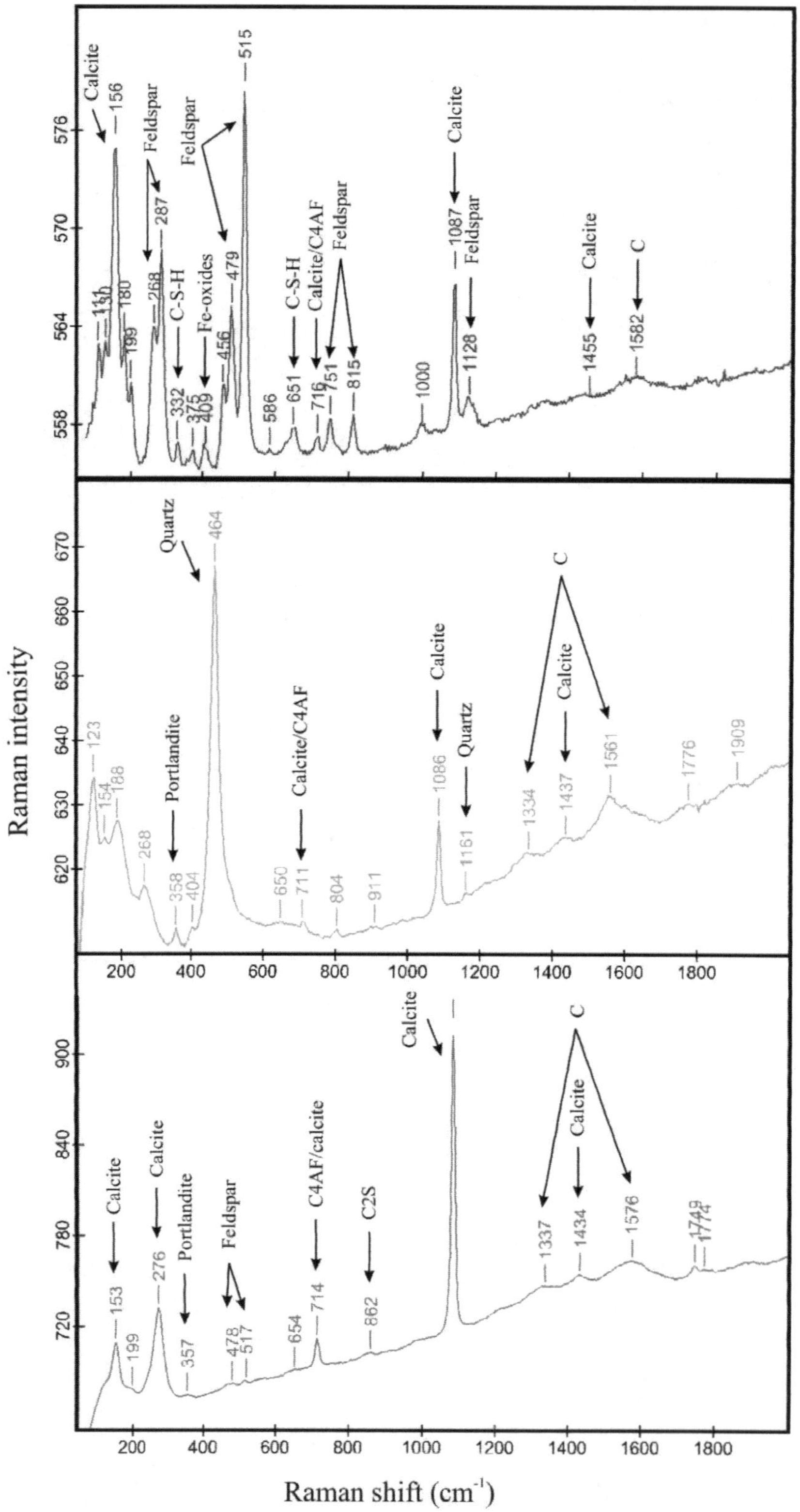


**Figure 3.** (continued).

*3.2.2. Sample VPC-1.* The boundaries quartz-cement paste and fragment of granitoid-cement paste (figure 4a) were selected for study. Cluster analysis (figure 4b, c) and optical profilometry (figure 4d) were performed on the sample. In the Raman image, the ITZ isn´t as distinct as in the OPC-1 sample. Average Raman spectra for individual clusters in concrete VPC-1 are shown on figure 5. As in the case of the sample OPC-1, C-S-H is a major hydration product of Portland cement and is identified by bands at 331–334 and 638–646 $cm^{-1}$. The bands at 357–360 $cm^{-1}$ correspond to the Ca-O bond vibration in portlandite, which was detected in three clusters, but its intensity is weaker than that of the OPC-1 sample. It means that only small amount of this phase is formed in VPC-1. Weakly intense band of ettringite at 980 and 986 $cm^{-1}$ was detected in two clusters. A thin layer (1–5 μm) of calcite is formed around quartz grains (figure 4c), this can significantly reduce the strength of this zone.

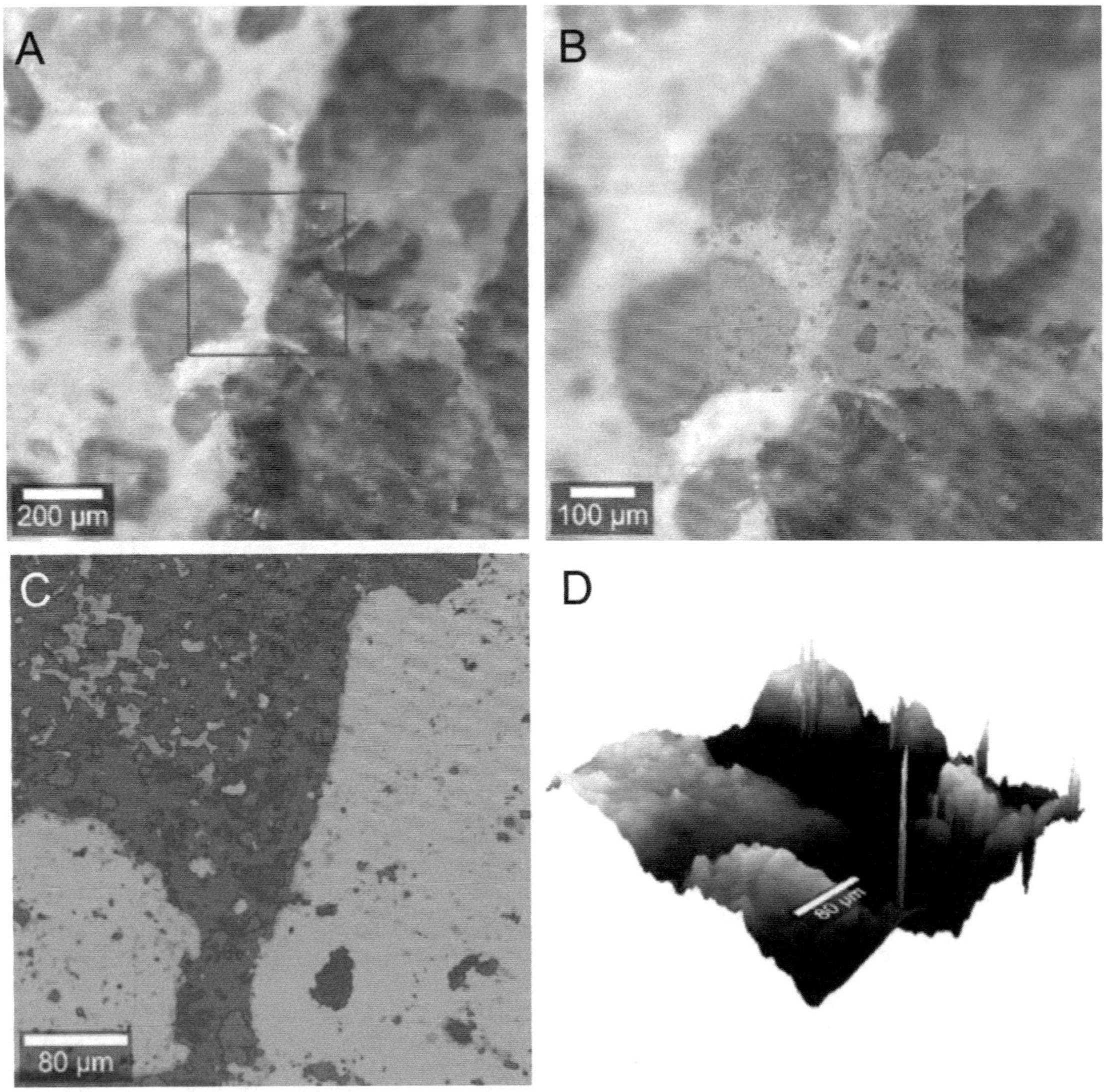


**Figure 4.** Sample VPC-1 **A)** Bright-field image under the confocal optical microscope; **B**) Bright-field image of the sample underlain by Raman image in the studied area; **C)** Raman image of studied area of the sample; **D)** Sample profile analysed by optical profilometry.

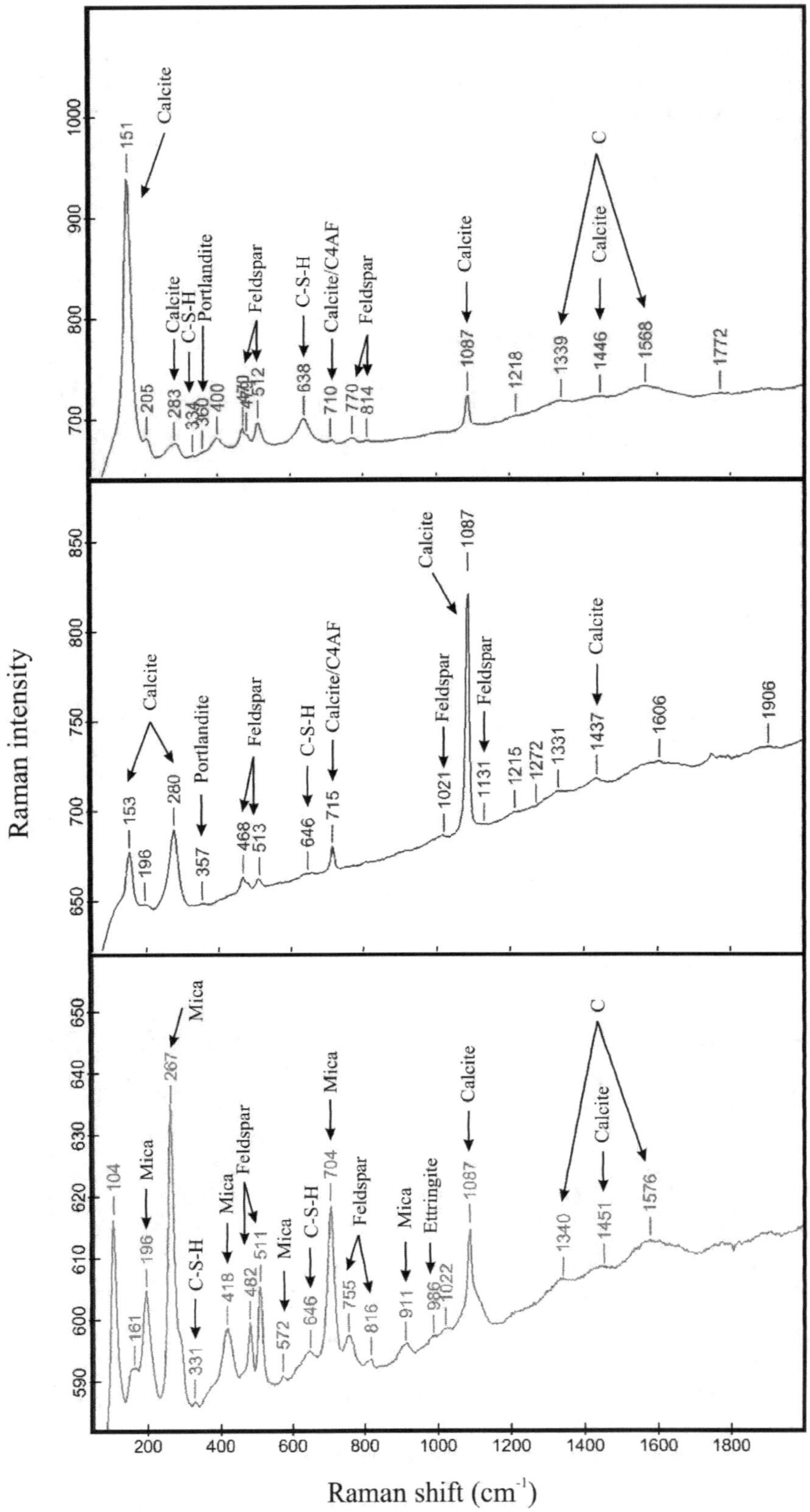


**Figure 5.** Raman spectra for individual clusters from sample VPC-1.

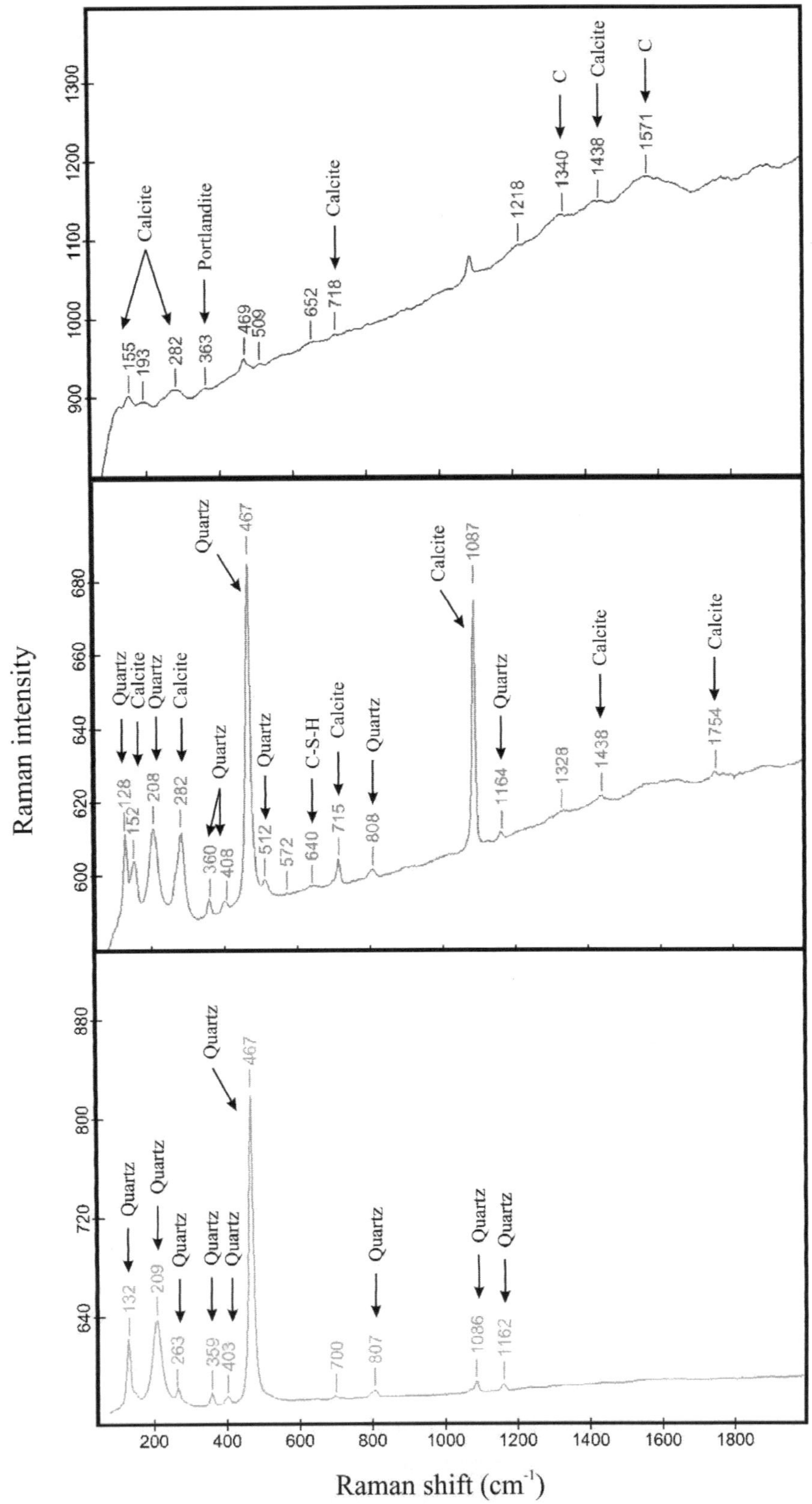


**Figure 5.** (continued).

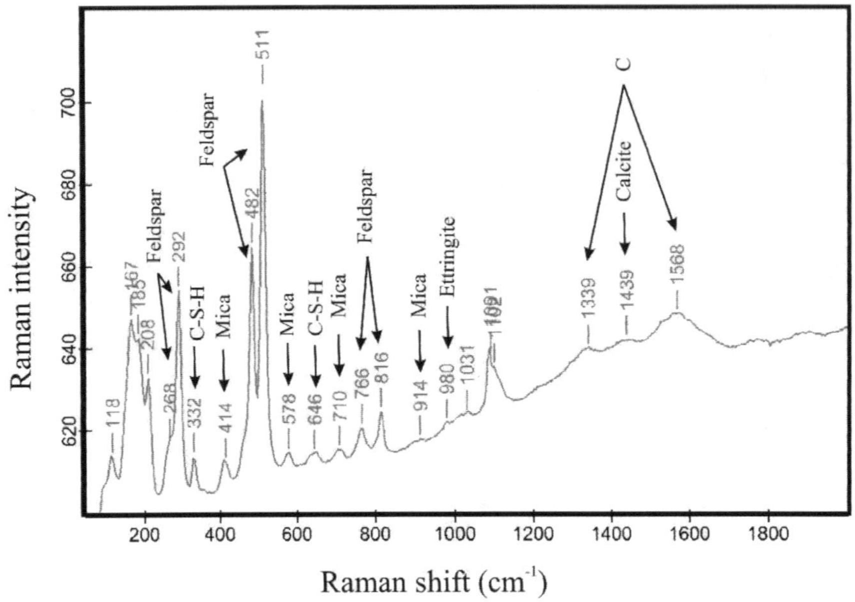


**Figure 5.** (continued).

## 4. Discussion

Concrete doped with limestone powder and metakaolin has undoubtedly advantages in comparison with concrete, where the binder is only ordinary Portland cement. The ITZ in this concrete has lower thickness and contains fewer undesirable phases, mainly portlandite. Although this concrete has less carbonation resistance and a higher amount of calcite accumulates in the ITZ. Especially calcite forms layer around quartz, it causes that the strength of this zone may be lower.

Engineering properties of LC3 cement were studied by [e.g. 15, 16, 17, 18, 19], however it is not possible to compare the data of cements and the data of concrete prepared in this work, since the addition of the aggregate significantly affects the mechanical properties, mainly the strength characteristics.

Mechanical properties of concrete with LC3 cement were studied by fewer authors [e.g. 20, 21, 22]. In the research by Nguyen et al. [21] basalt was used as coarse-grained aggregate and samples were cured in lime-saturated water bath for 7 days at room temperature. Subsequently, the samples were stored for 21 days at room temperature and 50% relative humidity. In contrast, granitoid rock was used as coarse-grained aggregate in our research and only distilled water was used in curing for 28 days. Compressive strength of concrete with LC3 in the study by Nguyen et al. [21] reached up to 55 MPa after 28 days and thus had greater compressive strength than the reference sample. In contrast, in the case of our prepared samples, the concrete with the addition of limestone powder and metakaolin had lower strength than the reference sample (see table 2).

As in our research, also in the research of Dhandapani et al. [22] granite was used as coarse aggregate. Compared to our research, in the study presented in [21] gypsum and the PCE superplasticizer were added to the mixture. It was found that concrete with LC3 binder has comparable compressive strength to concrete with conventional Portland cement. Moreover, compared to concrete with OPC, concrete with LC3 has higher resistance to chloride penetration [22]. Compared to our data, concrete samples from the work of Dhandapani et al. [22] have higher compressive strength.

Raman spectroscopy in comparison with other methods (XRD, WDX) represents a non-destructive research method. The laser beam is directed at a specific point in the sample and is also possible to analyse different layers of the sample, which is great advantage of this method in comparison with e.g. XRD [23]. It is possible to analyse aqueous solutions or gaseous phases that can be closed in pores of concrete using Raman spectroscopy [23, 24]. Amorphous phases can also be analysed using Raman spectroscopy [25, 26], as was also verified in this work, when C-S-H gel was identified in ITZ (figure 3 and 5). This method is used to study the hydration of cement paste in situ [27, 28]. The problem in the study of OPC using Raman spectroscopy is the background fluorescence signals and weak Raman signals [29, 30], which can also be seen in the spectra of samples in our work (figure 3 and 5). Fluorescence in OPC is most often caused by organic impurities or contamination during grinding and polishing of the samples [31, 32]. Raman bands of $C_3S$ and $C_2S$ in OPC are in many cases shifted against pure standards [29, 33]. From this perspective, Raman spectroscopy is more suitable for studying mixtures without OPC, e.g. sulfoaluminate cements [34] or lime mortars [35].

## 5. Conclusion

Raman spectroscopy and mechanics properties study of concrete with ordinary Portland cement and concrete with the addition of metakaolin and limestone powder were presented. The results show the following:

- C-S-H is the main hydration product of Portland cement in both samples.
- In concrete with ordinary Portland cement, the interfacial transition zone has higher thickness and more portlandite is formed in this zone.
- In concrete with the addition of metakaolin and limestone powder, a thin rim of calcite is formed around quartz particles.
- Concrete OPC-1 has higher compressive and tensile strength and lower density and absorbability compared to concrete VPC-1.

Further research will be focused on the study of interfacial transition zone using electron microprobe and deepening the study using Raman spectroscopy on other concrete samples with the addition of supplementary cementitious materials.

**Acknowledgments**

This work was created with the financial support of the Ministry of Industry and Trade within the institutional financing of long-term conceptual development of a research organization (Research Institute of Building Materials), the MEYS CR (Large RI Project LM2018129 Czech-BioImaging) and ERDF (project No. CZ.02.1.01/0.0/0.0/16_013/0001775).

## 6. References

[1] Duan P, Shui Z, Chen W and Shen Ch 2013 Enhancing microstructure and durability of concrete from ground granulated blast furnace slag and metakaolin as cement replacement materials *J. Mater. Res. Technol.* **2** 1 pp 52–59

[2] Simčič T 2015 Influenece of supplementary cementitious materials on hydration, microstructure development and durability of concrete *Doctoral thesis* (Univerza v Ljubljani) p 117

[3] Wescott R F, McNulty M W, VanGeem M G and Gajda J 2010 Prospects for Expanding the Use of Supplementary Cementitious Materials in California *Coalition for Sustainable Cement Manufacturing & Environment* p 27

[4] Scrivener K L, Martirena F, Bishnoi S and Maity S 2018 Calcined clay limestone cements ($LC^3$) *Cem. Conc. Res.* **114** pp 49–56

[5] Khan M S H, Nguyen Q D and Castel A 2019 Carbonation of limestone calcined clay cement concrete *Martirena F et al. (eds.) Calcined Clays for Sustainable Concrete*, RILEM Bookseries 16 pp 238–243

[6] Darwin D 1995 The interfacial transition zone: "direct" evidence on compressive response *Mater. Res. Soc. Symp. Proc.* **370** pp 419–427

[7] Nili M and Ehsani A 2015 Investigating the effect of the cement paste and transition zone on strength development of concrete containing nanosilica and silica fume *Mat. Des.* **75** pp 174–183

[8] Scrivener K, Crumbie A K and Laugesen P 2004 The interfacial transition zone (ITZ) between cement paste and aggregate in concrete *J. Coll. Inter. Sci.* **12** 4 pp 411–421

[9] Wenk H R, Monteiro P J M, Kunz M, Chen K, Tamura N, Lutterotti L and Delacroz J 2009

Preferred orientation of ettringite in concrete fractures *J. Appl. Cryst.* **42** pp 429–432
[10] Wu K 2014 Experimental study on the influence of ITZ on the durability of concrete made with different kinds of blended materials *Doctoral thesis* (Ghent University) p 290
[11] Machovič V, Kopecký J, Němeček J, Kolář F, Svítilová J, Bittnar Z and Andertová J 2008 Raman micro-spectroscopy mapping and microstructural and micromechanical study of interfacial transition zone in concrete reinforced by poly(ethylene terephthalate) fibres *Ceram.-Silik.* **52** pp 54–60
[12] Machovič V, Lapčák L, Borecká L, Lhotka M, Andertová J, Kopecký L and Mišková L 2013 Microstructure of interfacial transition zone between PET fibre and cement paste *Acta Geodyn. Geomat.* **10** 1 169 pp 121–127
[13] Machovič V, Kolář F, Procházka P P, Pešková Š and Kuklík P 2006 Raman spectroscopy study of interfacial transition zone in cement composite reinforced by PP/PE and basalt fibres *Acta Geodym. Geomat.* **3** 2 pp 63–67
[14] Al-Bayati H, Tighe S and Baaj H 2016 Effect of different Treatment Methods on the Interfacial Transition Zone Microstructure to Coarse Recycled Concrete Aggregate *Conference of the Transportation of Canada* (Toronto) pp 1–29
[15] Tironi A, Scian A N and Irassar E F 2017 Blended Cements with Limestone Filler and Kaolinitic Calcined Clay: Filler and Pozzolanic Effects *J. Mater. Civ. Eng.* **29** 9
[16] Ferreiro S, Canut M M C, Lund J and Herfort D 2019 Influence of fineness of raw clay and calcination temperature on the performance of calcined clay-limestone blended cements *Appl. Clay Sci.* **169** pp 81–90
[17] Scrivener K, Avet F, Maraghechi H, Zunino F and Ston J 2019 Impacting factors and properties of Limestone Calcined Clay Cements (LC3) *Green Mater.* **7** 1 pp 3–14
[18] Antoni M, Rossen J E, Hernandez J F M and Scrivener K 2012 Cement Substitution by a Combination of Metakaolin and Limestone *Cem. Conc. Res.* **42** 12 pp 1579–1589
[19] Apsa and Ranga Rao V (2019) Performance of Limestone Calcined Clay Cement (Lc3). Int. J. Recent *Technol. Eng.* **7** 6C2 pp 260–265
[20] Avet F, Sofia L and Scrivener K 2019 Concrete Performance of Limestone Calcined Clay Cement (LC3) Compared with Conventional Cements *Adv. Civ. Eng. Mat.* **8** 3 pp 275–286
[21] Nguyen Q D, Khan M S H and Castel A 2018 Engineering Properties of Limestone Calcined Clay Concrete *J. Adv. Concr. Technol.* **16** pp 343–357
[22] Dhandapani Y, Sakthivel T, Santhanam M, Gettu R and Pillai R G 2018 Mechanical properties and durability performance of concretes with Limestone Calcined Clay Cement ($LC^3$) *Cem. Conc. Res.* **107** pp 136–151
[23] Smith E and Dent G 2005 Modern Raman Spectroscopy – A Practical Approach (John Wiley & Sons, Ltd) pp 224
[24] Chen W, Du R-G, Ye Ch, Zhu Y-F and Lin Ch 2010 Study on the corrosion behavior of reinforcing steel in simulated concrete pore solutions using in situ Raman spectroscopy assisted by electrochemical techniques *Electrochim. Acta* **55** 20 pp 5677–5682
[25] Garbev K, Stemmermann P, Black L, Breen Ch, Yarwood J and Gasharova B 2007 Structural Features of C–S–H(I) and Its Carbonation in Air—A Raman Spectroscopic Study. Part I: Fresh Phases *J. Am. Ceram. Soc.* **90** 3 pp 900–907
[26] Kirkpatrick R J, Yarger J L, McMillan P F, Yu P and Cong X D 1997 Raman spectroscopy of C-S-H, tobermorite and jennite *Adv. Cem. Bas. Mat.* **5** 3–4 pp 93–99
[27] Black L, Breen C, Yarwood J, Phipps J and Maitland G 2006 In situ Raman analysis of hydrating C3A and C4AF pastes in presence and absence of sulphate *Adv. Appl. Ceramics* **105** 4 pp 209–216
[28] Richardson I G, Skibsted J, Black L and Kirkpatrick R J 2010 Characterisation of cement hydrate phases by TEM, NMR and Raman spectroscopy *Adv. Cem. Res.* **22** 4 pp 233–248
[29] Martínez-Ramírez S and Fernández-Carrasco L F 2011 Raman spectroscopy: Application to cementitious systems *Construscion and Building: Desing, Materials, and Techniques* pp 233–

244
[30] Newman S P, Clifford S J, Coveney P V, Gupta V, Blanchard J D, Sefarín F, Ben-Amotz D and Diamond S 2005 Anomalous fluorescence in near-infrared Raman spectroscopy of cementitious materials *Cem. Conc. Res.* **35** 8 pp 1620–1628
[31] Chris D and Dyer P J H 1993 The Raman spectroscopy of cement minerals under 1064 mn excitation *Spectrochimica Acta* **49A** 5–6 pp 715–722
[32] Black L 2011 Raman spectroscopy of cements. Problems and possibilities. In: Mauko E, Kosec T, Kopar T, Gartner N, (eds.) *Proceedings of 13th Euroseminar of Microscopy Applied to Building Materials. EMABM - 13th Euroseminar of Microscopy Applied to Building Materials, 14 Jun 2011 - 18 Jun 2011, Ljubljana, Slovenia* pp 6–7
[33] Black L 2009 Raman spectroscopy of cementitious materials *Spectrosc. Inorg. Compounds* **40** pp 72–127
[34] Torréns-Martín D, Fernández-Carrasco L and Martínez-Ramírez S 2013 Hydration of calcium aluminates and calcium sulfoaluminate studied by Raman spectroscopy *Cem. Conc. Res.* **47** pp 43–50
[35] da Silva Dalto D P, da Conceição Ribeiro R C and de Moura L C R 2018 Characterization of the Lime Mortars of the Rui Barbosa House Museum in Rio De Janeiro, Brazil *Minerals* **8** 50 p 19

# Negative effect of zinc compounds on hydration kinetics of ordinary Portland cement

**L Matejka[1], P Siler[1], R Novotny[1], J Svec[1], J Masilko[1] and F Soukal[1]**

[1] Brno University of Technology, Faculty of Chemistry, Institute of Materials Chemistry, Purkynova 464/118, 612 00 Brno, Czech Republic

Email: xcmatejkal@vutbr.cz

**Abstract.** Metals and metal containing compounds have significant effect on cement hydration mechanism and kinetics. The recycling of waste materials leads to increase of content of metals in various industrial products and visibly affects properties of the Portland cement. One of the metals that has been known to drastically increase the setting time is zinc. Zinc forms various hydroxyl species with calcium cations. Pore solution thus has lowered concentration of calcium which suppresses formation of portlandite. Due to insufficient concentration of calcium the formation of CSH gel and further hydration of clinker grains is inhibited The aim of this study is to quantify the negative effects of zinc at different concentrations at ambient temperature at 20°C. Isoperibolic calorimetry was used to measure the negative effects of zinc on hydration kinetics of ordinary Portland cement. Zinc was added to the cement in the form of two soluble salts $Zn(NO_3)_2$ hydrate, $ZnCl_2$ and a poorly soluble compound ZnO. The concentration of zinc added was chosen between 0.1 and 1 wt. % of clinker. Significant retardation of cement hydration with increasing zinc content was proven.

## 1. Introduction

Portland cement is a hydraulic binder produced by burning of ground limestone, clay and other raw materials in rotary kiln. Created clinker is ground up and mixed with small amounts of gypsum. Produced clinker consist of many phases. The main clinker phases are tricalcium silicate (abbr. $C_3S$–$3CaO\cdot SiO_2$ or $Ca_3SiO_5$, alite), dicalcium silicate (abbr. $C_2S$–$2CaO\cdot SiO_2$ or $Ca_2SiO_4$, belite), tricalcium aluminate (abbr. $C_3A$–$3CaO\cdot Al_2O_3$ or $Ca_3Al_2O_6$, celite) and calcium aluminoferrite (abbr. $C_4AF$–$4CaO\cdot Al_2O_3\cdot Fe_2O_3$ or $Ca_4Al_2Fe_2O_{10}$, brownmillerite) [1].

Various metals can be introduced into cement in all stages of production or application. Limestone based on the specific deposit tends to contain other minerals such as sphalerite (ZnS). Ordinary Portland cement is also widely researched as a binder for solidification and immobilization of toxic waste containing metals. Solidification and immobilization of toxic elements encompasses reducing bioavailability, leachability, and mobility of hazardous elements [1–5].

Zinc can be present in admixtures, in its own separate phase in clinker or in solid solution with main clinker phases. Solid solution with $C_3S$ seems to be preferred over $C_2S$ [6]. Ions of zinc present in pore solution hydrolyse into various hydroxyl species. These compounds coat hydrating grains of cement clinker and react with calcium cations evolved into the pore solution during formation of CSH gel. The resulting insoluble $Ca(Zn(OH)_3)_2\cdot 2\,H_2O$ depletes concentration of $Ca^{2+}$ preventing the formation of portlandite and severely suppressing saturation of pore solution and further hydration of cement grains. Precipitation of portlandite and CSH gel is inhibited until all zinc has been converted into insoluble $Ca(Zn(OH)_3)_2·2H_2O$. Once this process reaches completion hydration continues once more. Reactions leading to $Ca(Zn(OH)_3)_2·2\ H_2O$ are shown in equations (1) and (2) [7–10].

$$ZnO + H_2O + 2\ OH^- \rightarrow Zn(OH)_4^{2-} \quad (1)$$
$$2\ Zn(OH)_4^{2-} + Ca^{2+} + 2\ H_2O \rightarrow Ca(Zn(OH)_3)_2 \cdot 2\ H_2O + 2\ OH^- \quad (2)$$

The purpose of this study is to show the effect of zinc compounds $ZnCl_2$, $Zn(NO_3)·6\ H_2O$ and poorly soluble compound ZnO at different concentrations on the hydration induction period of ordinary Portland cement at lower ambient temperature than most studies and compare results with studies using higher ambient temperature. Exposure of zinc oxide to humidity affects both solubility and particle size distribution. Moisture present in chemicals can lead to misleading results by altering the concentration of zinc in the zinc source. These parameters can also affect reaction kinetics of zinc oxide in pore solution. Monitoring of these effects is done using isoperibolic calorimetry. It is apparent that ambient temperature has significant influence on both hydration of cement and mechanisms behind zinc induced prolonging of induction periods.

## 2. Experimental

### *2.1. Materials and Sample Preparations*

All samples were prepared using cement CEM I 42,5 R produced by Mokrá—Českomoravský cement, a.s., Heidelberg Cement Czech Republic. Chemical composition of this cement is provided in technical documentation and is shown in table 1 [11]. Each cement sample contained a specific amount of zinc equal to weight % of cement (1, 0.5 and 0.1%) and 0.4 water to cement ration. Cement pastes were prepared using standard kitchen aid mixer according to the norm ČSN EN 196-1.

**Table 1**. Chemical composition of used cement [11].

| Component | Cement (wt. %) |
|---|---|
| CaO | 63.7 |
| $SiO_2$ | 19.6 |
| $Al_2O_3$ | 4.8 |
| $Fe_2O_3$ | 3.3 |
| MgO | 1.4 |
| $SO_3$ | 3.1 |
| $Cl^-$ | 0.040 |
| $K_2O$ | 0.75 |
| $Na_2O$ | 0.19 |
| Insoluble | 0.7 |
| LOI | 3.4 |

Zinc was introduced into cement paste in a form of soluble salts $ZnCl_2$ and $Zn(NO_3)_2·6\ H_2O$ and poorly soluble compound ZnO in concentrations 0.1; 0.5 and 1 wt. % of pure zinc in cement. The humidity of the oxide was determined with Kern MLS 50-D humidity analyser. Zinc oxide used as one of zinc sources contained on average 0.35 weight % of moisture. ZnO was used in stock and dried form. ZnO was dried at 80°C for 72 h. Zinc chloride and zinc nitrate hexahydrate were purchased from Lach-Ner, s.r.o. wile zinc oxide was purchased from Lachema, s.r.o..

### *2.2. Isoperibolic Calorimetry*

One of the best methods for evaluating hydration of cement is multicell isoperibolic calorimetry. This type of device allows monitoring of multiple samples and references allowing easy comparison of different cement mixtures. This method is widely used for the study of the effects of different admixtures such as superplasticizers, hardening accelerators and various secondary industrial products on setting of cement pastes [12].

Cement mixtures were weighted (300 g) into polystyrene cup with thermocouples and covered with thermo-insulating top. Calorimetric curves showing cement hydration and differences between various samples were obtained using isoperibolic calorimetry in a room with ambient temperature at 20°C.

## 3. Results and discussion

Mixtures of samples were prepared outside of the calorimeter thus the first peak on hydration curve was not measured in its entirety. Nevertheless, the early stages of cement hydration are not significant to this study and do not significantly contribute to any changes during hydration.

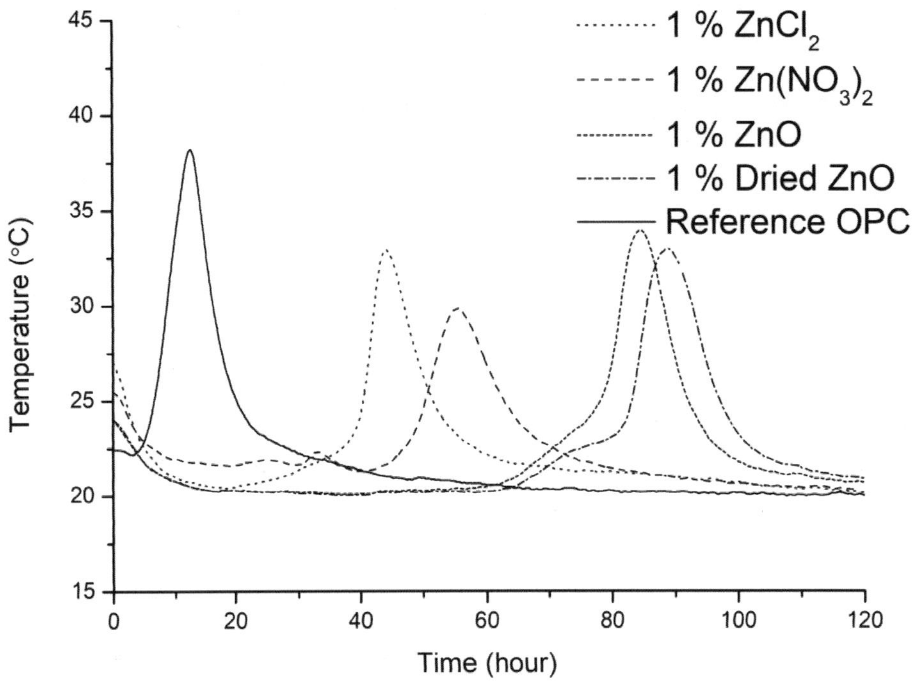


**Figure 1**. Isoperibolic calorimetry curves for samples with 1% weight concentration of zinc.

From calorimetric curves shown in figure 1 it is apparent that cement hydration of samples doped with zinc compounds was severely inhibited as peaks are shifted to the right of the reference peak. There also appears to be a small wide peak at around 25 h on a curve for sample with zinc nitrate. The heat evolution causing this peak was probably a result of dissolution of the zinc salt or crystallization on the surface of the cement grains as explained in works [13, 14].

Maximum temperatures reached by sample peaks were also lower than that of reference. Maximum temperatures reached by the samples were 32.9°C for cement with chloride, 29.9°C for nitrate, 34.0°C for untreated zinc oxide and 33.2°C for samples with dried zinc oxide. This is probably a result of zinc compounds slowing hydration of clinker grains.

At the same concentration of zinc, $ZnCl_2$ seems to show lower inhibitory ability than zinc nitrate. zinc oxide showed the highest inhibition of hydration. The difference in behaviour of samples doped with zinc oxide is due to lower solubility in water than zinc chloride and zinc nitrate. Smaller inhibition by chloride and nitrate is due to the ability of chloride and nitrate ions to accelerate cement hydration with chloride being more effective accelerator at this concentration [15].

When compared with similar studies done by Šiler et al. [16, 17] who used 25°C (as opposed to 20°C in this study) as ambient temperature, first noticeable differences are wider hydration peaks with higher maximum temperature when higher ambient temperature is used. This corresponds with accelerated hydration of cement and formation of hydration products. Samples containing chlorides show comparable induction period with samples in this study but had reached significantly higher maximum temperature at around 37 and in contrast with 32.9°C for our samples. Maximum temperatures of samples with zinc oxide are comparable with samples in this study but the length of induction period was different. This could be explained by lower solubility of dried zinc oxide, but the difference does not seem to be very significant. In study published by Kolářová et al. [18] these cement systems were studied using isothermal calorimetry. Induction period measured by this method was at around 80 h which is very similar to 77.8 h with dried zinc oxide and 73.3 h with untreated zinc oxide. Summary of determined induction periods is shown in table 2.

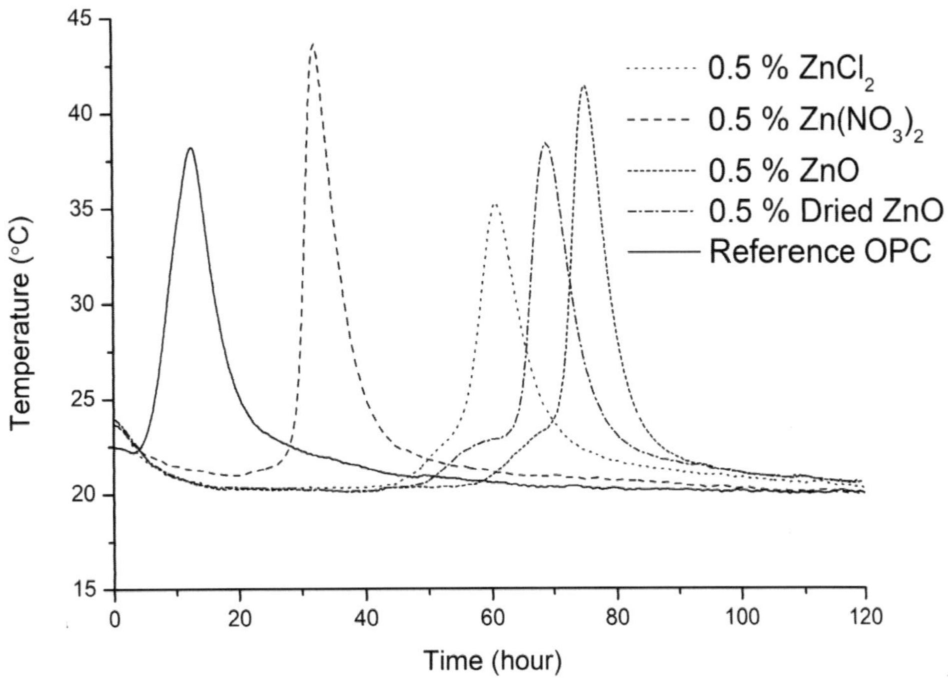


**Figure 2.** Isoperibolic calorimetry curves for samples with 0.5% weight concentration of zinc.

Behaviour of samples with concentration of zinc equal to 0.5% of weight of the clinker visibly changed as is demonstrated in figure 2. Peaks belonging to the zinc oxide samples shifted to the left indicating shorter induction period. Both zinc oxide peaks were thinner than those of samples with concentration of 1%. This indicates faster hydration relative to the previous samples.

Peaks of chloride and nitrate samples were visibly thinner than those of previous samples. The maximum temperature reached by nitrate sample was higher than maximum temperature of the reference. This indicates accelerated hydration in comparison to the samples with higher concentration of zinc. Significant difference was observed in induction periods of nitrate and chloride samples. Overall lengths of induction periods of all samples are in table 2. While the induction period of nitrate sample dropped from 48.2 h to 29 h with decrease in concentration of zinc, the induction period of chloride sample increased from 38.9 h to 56 h with decrease in concentration of zinc. This could be result of chloride and nitrate anions accelerating hydration through different mechanisms. While chloride anions are said to affect hydration by incorporation into CSH gel, disrupting it, allowing more water to be available to hydration reactions, nitrates affect formation of ettringite and the hydration of β-$C_2S$. Next parameter to consider is solubility of various ion species in pore solution with differing concentrations of chlorides and nitrates in such complicated systems [19, 20].

Results of Šiler [16, 17] measurements corresponding to 0.5% chloride show higher maximum temperature in comparison with 1% chloride, while the induction period was similar. In contrast with measurements in this study where induction period increased. This could be explained by loss of inhibitory effect with decreasing zinc ion concentration, loss of acceleration by chloride ions but with added acceleration caused by higher ambient temperature. Samples doped with nitrates measured by Šiler show significant increase in maximum temperature with decrease of concentration of nitrate. This correlates with presented results. Although sample of 0.5% nitrate measured at higher temperature showed shorter induction period than our results, the decrease in induction period with decreasing concentration was not as high as in our experiment. [16, 17].

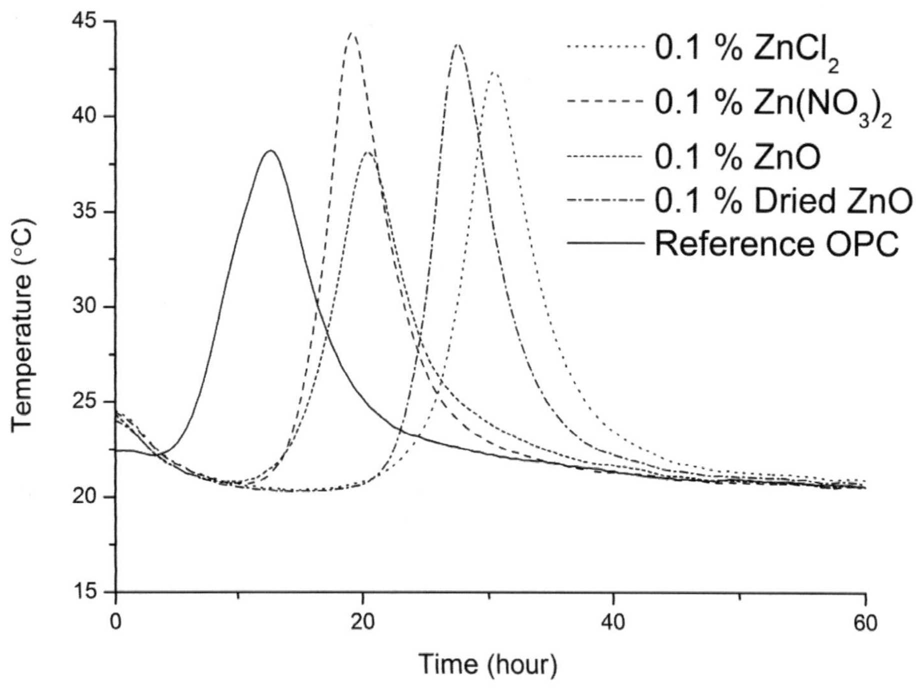


**Figure 3**. Isoperibolic calorimetry curves for samples with 0.1% weight concentration of zinc.

**Table 2**. Summary of induction periods measured based on isoperibolic curves.

| | Induction period (hour) | | | | |
|---|---|---|---|---|---|
| Concentration of zinc (wt.%) | $ZnCl_2$ | $Zn(NO_3)_2$ | ZnO untreated | ZnO dried | Cement |
| 0.0 | | | | | 4.9 |
| 0.1 | 27.1 | 13.7 | 14.9 | 22.6 | |
| 0.5 | 56.0 | 29.0 | 63.5 | 56.5 | |
| 1.0 | 38.9 | 48.2 | 73.3 | 77.8 | |

Differences between lengths of induction periods and maximum temperatures reached decrease with decreasing concentration of zinc. Peaks of all samples shifted to the left indicating shorter induction periods as seen in figure 3. Samples doped with zinc oxide reached higher maximum temperature than the reference cement. Induction periods of all samples ranged between 13.7 h and 27.1 h. These are significantly longer induction periods than the induction period of reference sample. These results are comparable with literature [16, 17].

Induction periods measured by Kolářová [18] ranged between 10 and 20 h. Sample doped with zinc chloride had significantly longer induction period than those doped with zinc nitrate. This phenomenon correlates with results of Kolářová. Samples doped with zinc oxide had the longest induction period out of her measurements. These results are very similar to our measurement with untreated zinc oxide samples. Our samples of cement with dried zinc oxide show significantly longer induction period than samples with untreated oxide.

## 4. Conclusion

The aim of this study was to monitor the effects of various zinc compounds on induction period of cement hydration at 20°C ambient temperature as opposed to 25°C used in other studies [16–18]. Compounds used as zinc source in this study were soluble salts $ZnCl_2$ a $Zn(NO_3)·6\ H_2O$ and insoluble compound ZnO in dried and not dried form.

- The length of induction period was increasing with increasing concentration of specific zinc source except for sample with 1 weight % of zinc in a form of $ZnCl_2$ due to acceleration effect of chloride anions at this concentration.
- Ambient temperature of 20°C was used during measurements (as opposed to of 25°C as used in studies [16–18]) which resulted in significantly higher peak temperatures reached at high concentrations of zinc.
- Differences in the length of induction periods between zinc chloride and zinc nitrate samples are bigger at 20°C ambient temperature as opposed to 25°C used in studies [16–18].
- The difference between induction periods and maximum temperatures reached by cement samples with dried and untreated zinc oxide was relatively small. Induction period of both samples were similar with consistent difference between 5 to 7 h. Untreated zinc oxide contained 0.35 weight % of water. This is insignificant amount that probably would not influence the results. The difference in hydration induction periods was probably caused by different particle size distribution of heat treated and untreated zinc oxide and different level of homogenization of sample pastes.

## Acknowledgment

This research has been achieved with the financial support by the project: GA19-16646S "The elimination of the negative impact of zinc in Portland cement by accelerating concrete admixtures", with financial support from the Czech Science Foundation.

## References

[1] Lea F and HewlettT P 1998 *Lea's chemistry of cement and concrete* 4th ed. (New York: Copublished in North, Central, and South America) by J. Wiley ISBN 03-405-6589-6

[2] Niu M, Guoxin L I, Wang Y, LI Q, Han L and Song Z 2018 Comparative study of immobilization and mechanical properties of sulfoaluminate cement and ordinary Portland cement with different heavy metals *Construction and Building Materials* vol. **193** pp 332–343 DOI: 10.1016/j.conbuildmat.2018.10.206 ISSN 09500618

[3] Wang Y-S, Dai J-G, Wang L, Tsang D and Poon Ch 2018 Influence of lead on stabilization/solidification by ordinary Portland cement and magnesium phosphate cement *Chemosphere* vol. **190** pp 90–96 DOI: 10.1016/j.chemosphere.2017.09.114 ISSN 00456535

[4] Rachman R, Bahri A and Trihadiningrum Y 2018 Stabilization and solidification of tailings from a traditional gold mine using Portland cement *Environmental Engineering Research* vol. **23**(2) pp 189–194 DOI: 10.4491/eer.2017.104 ISSN 1226-1025

[5] Chen L, Wang L and Cho D-G et al. 2019 Sustainable stabilization/solidification of municipal solid waste incinerator fly ash by incorporation of green materials *Journal of Cleaner Production* vol. **222**(222) pp 335–343 DOI: 10.1016/j.jclepro.2019.03.057 ISSN 09596526

[6] Andrade F R D, Maringolo V and Kihara Y 2003 Incorporation of V, Zn and Pb into the crystalline phases of Portland clinker *Cement and Concrete Research* vol. **33**(1) pp 63–71 DOI: 10.1016/S0008-8846(02)00928-6 ISSN 00088846

[7] Gineys N, Aouad G and Damidot D 2010 Managing trace elements in Portland cement – Part I: Interactions between cement paste and heavy metals added during mixing as soluble salts *Cement and Concrete Composites* vol. **32**(8) pp 563–570 DOI: 10.1016/j.cemconcomp.2010.06.002 ISSN 09589465

[8] Gineys N, Aoud G and Damidot D 2011 Managing trace elements in Portland cement – Part II: Comparison of two methods to incorporate Zn in a cement *Cement and Concrete Composites* vol. 33 pp 629–636 DOI: 10.1016/j.cemconcomp.2011.03.008 ISSN 09589465

[9] Ataie F, Juenger M, Taylor-Lange S and Riding K 2015 Comparison of the retarding mechanisms of zinc oxide and sucrose on cement hydration and interactions with supplementary cementitious materials *Cement and Concrete Research* vol. **72** pp 128–136 DOI: 10.1016/j.cemconres.2015.02.023 ISSN 00088846

[10] Fernández Olmo I, Chacon E and Irabien A 2001 Influence of lead, zinc, iron (III) and chromium (III) oxides on the setting time and strength development of Portland cement *Cement and Concrete Research* vol. **31**(8) pp 1213–1219 DOI: 10.1016/S0008-8846(01)00545-2 ISSN 00088846

[11] *CEM I 42,5 R Portlandský cement EN 197-1: Technický list* 2019 (Online: https://www.heidelbergcement.cz/cs/cement/volne-lozeny-cement/cemi425r)

[12] Brandštetr J, Polcer J, Krátký J, Holešinský R and Havlica J 2001 Possibilities of the use of isoperibolic calorimetry for assessing the hydration behavior of cementitious systems *Cement and Concrete Research* vol. **31**(6) pp 941–947 DOI: 10.1016/S0008-8846(01)00495-1 ISSN 00088846

[13] Rossetti V A and Medici F 1995 Inertization of toxic metals in cement matrices: effects on hydration, setting and hardening *Cement and Concrete Research* **25**(6) pp 1147–1152

[14] Weeks C, Hand R and Sharp J 2008 Retardation of cement hydration caused by heavy metals present in ISF slag used as aggregate *Cement and Concrete Composites* vol. **30**(10) pp 970–978 DOI: 10.1016/j.cemconcomp.2008.07.005 ISSN 09589465

[15] Myrdal R 2007 *Advanced cementing materials controlling hydration development: Accelerating admixtures for concrete* (Trondheim) Online: https://www.sintef.no/globalassets/sintef-byggforsk/coin/sintef-reports/sbf-bk-a07025_accelerating-admixtures-for-concrete.pdf

[16] Šiler P, Kolářová I, Másilko J, Novotný R. and Opravil T 2018 The Effect of Zinc on the Portland Cement Hydration *Key Engineering Materials* vol. **761** pp 131–134 DOI: 10.4028/www.scientific.net/KEM.761.131 ISSN 1662-9795

[17] Šiler P, Kolářová I, Másilko J, Novotný R., Opravil T, Pořízka J, Bednárek J and Švec J 2018 Application of isothermal and isoperibolic calorimetry to assess the effect of zinc on cement hydration *Journal of Thermal Analysis and Calorimetry* vol. **133**(1) pp 27–40 DOI: 10.1007/s10973-017-6815-1 ISSN 1388-6150

[18] Kolářová I, Šiler P and Šoukal F 2014 The Influence of Zinc on the Hydration and Compressive Strength of Portland Cement *Advanced Materials Research* vol. **1000** pp 43–46 DOI: 10.4028/www.scientific.net/AMR.1000.43 ISSN 1662-8985

[19] Abdelrazig B E I, Bonner D G, Nowell D V, Dransfield J M and Egan P J 1999 The solution chemistry and early hydration of ordinary portland cement pastes with and without admixtures

*Thermochimica Acta* **340-341** pp 417–430 DOI: 10.1016/S0040-6031(99)00286-5 ISSN 00406031

[20] Silva D 1999 *Impact of Accelerators and Retarders on the Hydration of Portland Cement: International Summit on Cement Hydration Kinetics* (Quebec) doi:10.1016/s0040-6031(99)00286-5

IOP Conf. Series: Materials Science and Engineering 1039 (2021) 012005 doi:10.1088/1757-899X/1039/1/012005

# Crushed bricks as aggregate in cement based binder

**J Palovcik**[1]**, T Opravil**[1]**, F Soukal**[1] **and J Masilko**[1]

[1] Brno University of Technology, Faculty of Chemistry, Institute of Materials Science, Purkynova 464/118, Brno Cz-612 00, Czech Republic

Email: xcpalovcik@fch.vut.cz

**Abstract**. The work deals with the impact of crushed brick on properties of Portland cement binder. The source of brick recycled materials is not only the demolition waste, but also the waste from brick production. One way to reduce the amount of this recyclate is to use it in cement-based composite. The advantages of using bricks as aggregate are their lower bulk density compared to conventional aggregates. Economic and ecological benefits are also important, because less natural sources need to be mined. First, the phase composition of used brick aggregate, its particle size and morphology were analysed. Test specimens with different content of brick recyclate were prepared and tested for the mechanical properties. Subsequently, the microstructure of prepared samples was examined using a scanning electron microscope. The phase composition of samples was analysed by X-ray diffraction analysis.

## 1. Introduction

The industry of building materials is largely dependent on natural raw materials which are not bottomless. One of the most commonly used materials is concrete, which consists of cement matrix, and aggregates. Cement matrix consist of Portland cement, water and other minorities [1]. The major part in most often represented concrete are fine and coarse aggregates 60–75% [2] which are largely obtained by mining. On the other hand, there is a lot of material that is produced from demolition or as a byproduct in the production of other building elements such as precast concrete or brick production. These materials can be sorted and processed so that they can be used as aggregates for concrete production. There are several motivations for reusing these materials. The first is the conservation of natural and economic resources. The second is solving the problem of waste accumulation. Last but not least, these are legal measures dealing with waste management.

Ceramics or fired brick were used as early as 3000 BC in early Indus Valley cities like Kalibangan [3]. Re-use of brick was common in Mediteranea by Romans, who used bricks as aggregates in Roman concrete (Opus caementum) [4].

This paper deals with the use of crushed bricks as type of fine aggregate replacing sand in cement-based concrete/composite. In a way, the paper is linked to previous paper: “Application of brick ground dust in systems based on Portland cement” which deals with partial replacement of cement by brick dust and addresses the influence of hydration of Portland cement by finely ground brick dust [5]. In the current article the amount of cement is constant and standard silica sand is replaced by brick recycled material.

The use of bricks as aggregates is investigated in many works in two ways. One way is adding of fine ground fraction as a micro-filler or possibly pozzolan. The other way is the application of brick aggregates into concrete in fractions of about 2 cm or larger without the fine fraction. Aggregates partially reduce the total bulk density but also mechanical properties of brick-concrete. This material is used as one type of aggregate in lightweight construction concrete. It is usually stated that concrete using fine brick aggregate up to 15% is comparable to normal concrete. [6] However, the porosity of the brick body must be kept in mind, which causes high absorption of water. High water absorption can have the benefit of "internal curing effect" [7, 8].

Despite of the benefits, the use of brick pulp also results in reduced strengths, increased water absorption causes increased demand for mixing water and increased difficulty in mixing. This paper deals with the study of the effect on mechanical properties using crushed bricks of fraction 2 mm or less, including a fine fraction of less than 63 microns. The fraction was mainly used because the fine particles are always formed during the grinding of bricks. The compressive strength of bricks is about 20 MPa [9]. Crushing of bricks results in not only a brittle fracture and size reduction but also in abrasion and formation of fine fractions.

## 2. Materials and methods

Crushed bricks were produced by grinding bricks which had unsuitable shape or were defected and unsaleable. Bricks were produced in plant Hevlín in South Moravia region (HELUZ). Other used material was standard silica sand defined by ČSN EN 196-1. Portland cement CEM I 42.5 R from Mokrá plant (Heidelberg cement group) was used as a binder. Tap water was used as mixing water. The mineralogical composition was analyzed by powder X-ray diffraction analysis (XRD, Empyrean, Panalytical). The phase composition of raw materials is shown in figure 1 and present phases with explanatory notes are written in table 1.

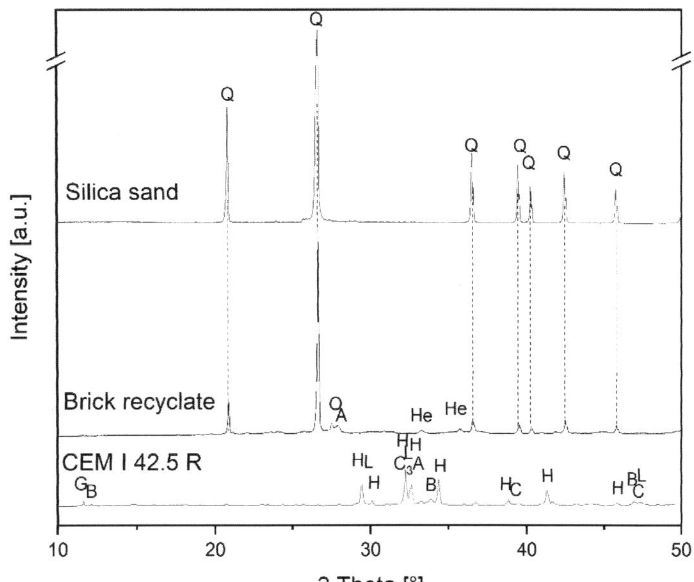


**Figure 1.** XRD analysis of used materials (Q – Quartz; O – Orthoclase; A – Albite; He – Hematite; G – Gypsum; B – Brownmillerite; H – Hatrurite; L – Larnite; C – Calcite).

**Table 1.** Phase composition of used materials.

| | | | | | | | |
|---|---|---|---|---|---|---|---|
| **Silica Sand** | Quartz (Q) | | | | | | |
| **Brick rec.** | Amorphous | | | | | | |
| **CEM I 42.5 R** | Hatrurite (H) | Quartz (Q) | Hematite (He) | Albite (Al) | Orthoclase (O) | Åkermanite | Muscovite |

The particle size distribution of crushed bricks recyclate was analyzed by six sieves with sieve size 2.0 mm, 1.0 mm, 0.5 mm, 0.25 mm, 0.125 mm, 0.063 mm and the results are shown in figure 2. The morphology of crushed bricks was observed by Scanning electron microscopy (SEM, Zeiss LS Evo 10), see figure 3.

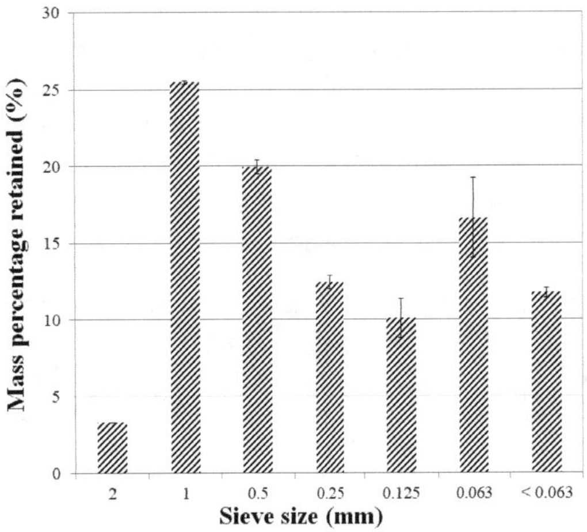


**Figure 2.** Particle size of crushed brick recyclate.

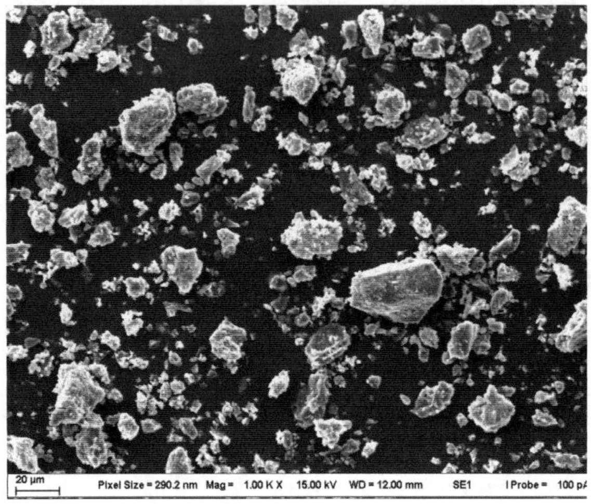


**Figure 3.** Morphology of crushed brick (SEM, mag. 10 000×).

The samples were prepared in such way that the standard sand was gradually replaced by crushed brick recyclate. As a reference the sample without the brick recyclate was prepared. The reference contains only normalized sand, cement and tap water. All samples were prepared according to the ČSN EN 196-1 standard.

The composition of prepared samples is shown in table 2. The water/cement ratio is marked w/c. Three test specimens (40 × 40 × 160 mm) were used for each testing of mechanical properties. The equipment for testing was DESTTEST and BETONSYSTEM. Prepared samples were cured for 1, 7 and 28 days in humidity chamber at laboratory temperature. The average values of flexural and compressive strengths are shown in figure 4 and figure 5. Parts of samples were milled and analyzed by XRD in order to determine the phase composition (figure 8). After 28 days the microstructure of sample was observed by SEM (figure 9 and figure 10).

**Table 2.** Composition of prepared samples.

| Sample | a | b | c | d | e | f |
|---|---|---|---|---|---|---|
| **crushed bricks [%]** | 0 | 20 | 40 | 60 | 80 | 100 |
| **crushed bricks [g]** | 0 | 270 | 1080 | 1620 | 2160 | 2700 |
| **sand [g]** | 2700 | 2430 | 1620 | 1080 | 540 | 0 |
| **sand [%]** | 100 | 80 | 60 | 40 | 20 | 0 |
| **CEM I 42,5 R [g]** | 900 | 900 | 900 | 900 | 900 | 900 |
| **water [g]** | 450 | 490 | 590 | 730 | 830 | 980 |
| **w/c** | 0.50 | 0.54 | 0.66 | 0.81 | 0.92 | 1.09 |

## 3. Results and discussion

### *3.1. Mechanical properties*

Flexural strength of prepared samples after 1 day decreased with rising amount of crushed bricks. After 7 and 28 days the strength of samples b, c and d was higher than that of reference sample a. It should be caused by angular shape of brick particles that arises during crushing. The surface of brick particles causes good bond between cement paste and brick aggregate which could increase the flexural strength performance [7]. As for compressive strength, the trend here is approximately similar. The reference sample a has about 3 MPa less compressive strength than sample b with 20% compensation of sand by crushed bricks after 28 days. Sample c with 40% compensation has only about 2 MPa less strength than the reference sample. The flexural tensile strengths are significantly lower after 28 days for the addition of 60% (sample d) and more. In the case of compressive strengths, there is a significant decrease of values for samples that contain more than 40% (c) of sand substitution with brick recycled material. The improvement of mechanical properties with the addition of brick aggregate is described in the literature rather rarely [10]. More often, a deterioration of the properties of about 15% to 22% by the brick pulp content is observed [6, 8, 11, 12]. The improvement of mechanical properties can be influenced by many factors. One of the most important factors is the use of pure brick recycled material compared to the use of pulp from demolition, which may contain impurities such as plaster, wood, clay and others. Another factor is the fine fraction content and the absence of aggregate greater than 2 mm. This can result in lower content of large defects in the brick aggregate and also reduce the potential for crack formation in the aggregate. On the other hand, with finer proportion of brick recycled material, its specific surface area and water absorption increase, thereby increasing the demand for mixing water content and the brick recycled material should be pre-saturated [13].

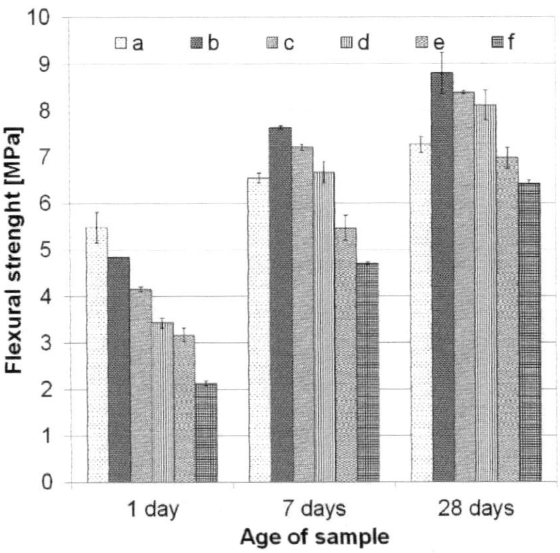


**Figure 4.** Development of flexural strength of prepared samples in time.

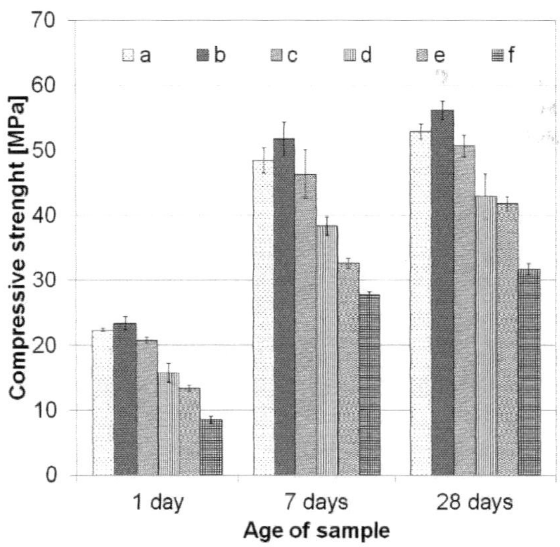


**Figure 5.** Development of compressive strength of prepared samples.

High water absorption can be an advantage. Water contained in the aggregate may later be provided for hydration, which may run to a higher degree. This phenomenon is described as the effect of internal curing effect and may be one of the reasons for achieving higher strengths [6].

Another effect may be the effect of the shape and surface of the brick aggregate, which is produced by crushing and promotes good mechanical bond between cement paste and aggregate.

The demonstration of good bonding of cement paste and aggregates is also the fracture of test specimens through brick aggregates (figure 7). But this is also given due to low mechanical strength of bricks (about 20 MPa depending on the type of bricks) [7].

In figure 6, significant reduction in the bulk density of prepared test specimens with the addition of brick aggregate can be observed. While maintaining the compressive strength of approximately 50 MPa, the bulk density can be reduced by 180 kg·m$^{-3}$ (sample c). When quartz sand is completely replaced, the density after 28 days is reduced by approximately 430 kg·m$^{-3}$. In figure 7, an increasing depth of carbonation can be observed with the increase of the amount of brick recycled material. This is probably due to its porosity, as well as the effect of volume density reduction.

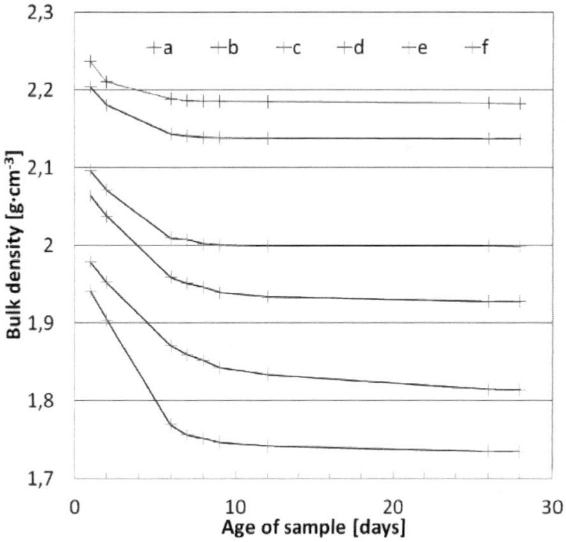


**Figure 6.** Development of bulk density in 28 days.

**Figure 7.** Carbonation line after 28 days (from the left top sample a to bottom left sample f) showed by phenolphthalein solution.

*3.2. Phase composition and microstructure*

In figure 8, the intense main diffraction peak of quartz corresponding to the plane (001) of 26.7° is omitted for better comparison of the intensities of other phases. The XRD analysis shows that replacing quartz sand with recycled brick in a Portland cement system does not give rise to any new crystalline phase. In the mixture b with 20% of the brick content, the quartz content increases. Quartz is the main phase of silica sand, but bricks also contain it. Furthermore, the intensity of quartz decreases as sand content is reduced. On the contrary, the content of other phases occurring in the brick (Albite, Orthoclase, Muscovite and Hematite) increases. The amount of unreacted clinker minerals (Larnite, Hatrurite) decreases with increasing content of brick recycled material, as does the amount of Portlandite. The principle is similar to that described by El-Didamony et all [14].

They used refractory bricks. These bricks contain phases formed at higher temperatures than ordinary bricks for building. The loss of clinker minerals may be due to better conditions of internal curing and filler effect [5, 6]. The loss of Portlandite with the addition of brick aggregate can be attributed to the pozzolan reaction [15]. On the other hand, the Calcite peaks can be observed and the reason of Portlandite reduction may be due to porous structure of brick aggregate and easier diffusion of $CO_2$ into the sample matrix.

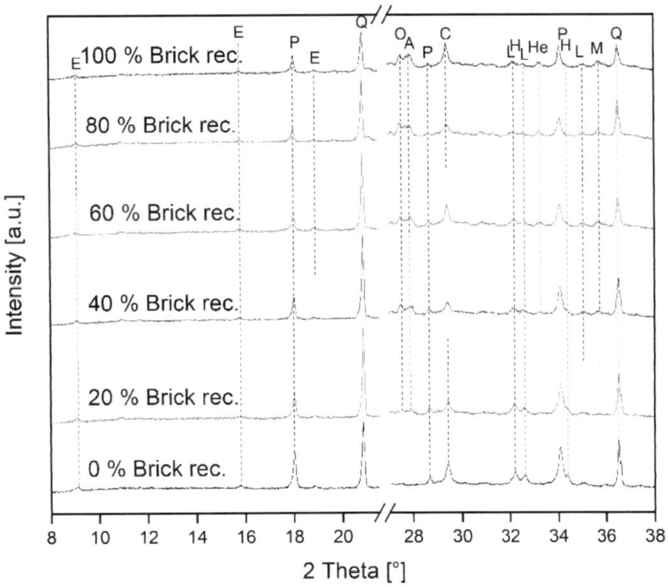


**Figure 8.** XRD analysis of prepared samples after 28 days (E – Ettringite; P – Portlandite, Q – Quartz; O – Orthoclase; A – Albite; C – Calcite; L – Larnite; H – Hatrurite; He – Hematite; M – Muscovite).

The microstructure of the samples was observed on the fracture area after 28 days in secondary electron mode at a magnification of 10 000×. The samples were gilded prior to observation. Only two samples were selected for the comparison in the paper – sample a (figure 9) without brick aggregate and sample f with 100% brick aggregate (figure 10) without sand. Both samples contain plate-like Portlandite crystals up to 1 micron thick. The sample containing brick aggregate exhibits finer CSH phase needle structure compared to sample a. The porosity of the sample is not comparable due to the fracture surface.

**Figure 9.** SEM image of sample a (0% brick agg.), magnification 10 000×.

**Figure 10.** SEM image of sample f (100% brick agg.), magnification 10 000×.

## 4. Conclusion

Partial replacement of quartz sand with recycled brick (up to about 40%) has positive effect on flexural and compressive strength. In addition, there is a decrease in bulk density. Brick aggregate has lower strengths than quartz sand, but compressive strengths of prepared samples with crushed bricks content are higher than reference without bricks. It means that the surface of brick aggregate probably reacts with the cement paste and contributes to the increase of strength. Or the fine fractions react with the cement as pozzolan or filler. Another positive feature is the so-called internal curing effect provided by the brick aggregate. The XRD results showed no new phases of hydration products. If any new phase was formed, it would be X-ray amorphous. The SEM results from the fracture surface suggest that the CSH phase formed in the case of brick aggregate is finer than in that with sand.

**Acknowledgement**

This work was supported by the project of specific research: FCH-S-19-5834 "Materiály a technologie pro 21. století " with the financial support of Brno University of Technology.

**References**

[1] Collepardi M 2009 *Moderní beton* (Praha: ČKAIT)
[2] Chen H, Yen T and Chen K *Cem. Conc. Res.* vol. **33**(1) pp 125–132
[3] Khan A and Lemmen C *Bricks and urbanism in the Indus Valley rise and decline*
[4] Římský beton 2017 *Antický svět* (Online: http://www.antickysvet.cz/26146n-beton)
[5] Palovčík J, Opravil T, Novotný R and Másilko J 2018 *IOP Conference Series: Materials Science and Engineering Institute of Physics Publishing* **379**(1)
[6] Cachim P 2009 *Constr. Build. Mater.* **23**(3) pp 1292–1297
[7] Debieb F and Kenai S 2008 *Constr. Build. Mater.* **22**(5) pp 886–897
[8] Poon Ch S and Chan D 2006 *Constr. Build. Mater.* **20**(8) pp 569–577
[9] Pytlík P 1995 *Cihlářství* (Brno: CERM)
[10] Akhtaruzzamana A and Hasnat A 1983 *Concr. Int.***2** pp 58–63
[11] Joyklad P, Nawaz A and Hussain Q 2018 *Suranaree Journal of Science and Technology* **25**(4) pp 349–362
[12] Aliabdo A, Abd-Elmoaty M and Hassan H 2014 *Alexandria Engineering Journal* **53**(1) pp 151–168
[13] De Brito J, Pereira A S and Correia J R 2005 *Cem. and Conc.Comp.* **27**(4) pp 429–436
[14] El-Didamony H, Amer A amd El-Hoseny S 2016 *Journal of Thermal Analysis and Calorimetry* **125**(1) pp 23–33
[15] Navrátilová E and Rovnaníková P 2016 *Constr. Build. Mater.* **120**(C) pp 530–539

IOP Conf. Series: Materials Science and Engineering 1039 (2021) 012006 doi:10.1088/1757-899X/1039/1/012006

# The external renders of the early 20th century architecture

**R Tislova[1], P Svoboda[2] and D Vsiansky[3]**

[1] Faculty of Restoration, University of Pardubice, Litomysl, Czech Republic
[2] National Heritage Institute, Methodological Center of Modern Architecture in Brno, Brno, Czech Republic
[3] Department of Geological Sciences, Faculty of Science, Masaryk University, Brno, Czech Republic

E-mail: renata.tislova@upce.cz

**Abstract**. External renders are an essential part of the architecture of the first half of the 20th century. The aesthetic effect of such plasterwork was achieved by surface treatment, colouring or by using various filler components that acted with their colour or structure. The binder of the plasters was most often Portland cement which played an important role in the processing of plasters or surface finishes, and in addition, formed the functional properties of the plasters (hardness, durability, resistance to external conditions).
This paper summarizes the results of a survey focused on the characterization of components used for the creation of external rendering taken from several architecturally significant buildings in the city of Brno dating back to the first half of the 20th century (villas, apartment buildings, public apartment houses). The survey into the collected samples performed by means of petrographic (light polarizing microscopy) and phase (powder X-ray diffraction) analysis revealed interesting information concerning the wide range of materials (components) used and the craftsmanship involved in their processing, all of which fundamentally influence the art-craft and utility properties of the buildings. The information gathered can serve as an important source of information for the renovation of early 20th century architecture.

## 1. External renders of the first half of the 20th century - introduction

In civil engineering, renders have always been the element of the façade that, to a certain extent dictates the aesthetics of the building itself as a whole. Throughout history, a wide range of materials have been used for renders, with the traditional binder being lime right up to the end of the 19$^{th}$ century. However, towards the end of the 19$^{th}$ century, the Secession period brought an unbelievable array of renders. The colourfulness and structure of the plasters were dictated not only by the pigments but also by the aggregates used and, above all, by the surface finishing technique. Plasters of varying compositions and surface structures could be applied on one building, with stucco framing geometrically unifying divergent surfaces. The demand for a unique expression motivated builders to experiment in developing plasters which would appear in a certain region and within a short period of construction. Successful experiments were further copied and adopted by other workshops, sometimes however, without knowledge of the original technique, thus increasing the number of varieties. The diversity of materials was further enriched by the use of Roman or Sorel cement and modified gypsum plaster binders, used mainly for making mortars and renders or for the creation of cast decoration elements integrated into the façades [1, 2, 4]. The variability of façade surfaces was not even affected by the dramatic increase in the production of Portland cement in the early 1900s and its increased use throughout the whole world as, among other applications, a binder in plasters. Cement binders therefore, can be found in all European centres on buildings in the style of Late-Classicism through individualist Modernism to Functionalism. The introduction of Portland cement into the making of plaster did not however force out lime. On buildings from the first half of the 20th century it is common to find "hard" cement-based renders alongside "soft" lime-based renders.

Various renders can coexist also as multi-strata layers; period instruction manuals of the period for masons generally recommend applying the render on exterior walls in two layers – the rough bottom (core) layer with a cement binder and sand mixed in a high ratio of from 1:3 to 1:5, followed by the second layer with its increased cement content because, as some manuals state, the aggregate lowers the binding properties of cement. Only some of the manuals recommend adding lime in order to slow the setting and improve the workability, but on the other hand the reason being the resulting poor resistance to moisture in exterior conditions [3–6].

Cement-based renders were carried out using an exceptionally wide range of technologies and craft techniques. Renders were created in such a way as to create an effect with their own distinct colour and inner structure (pigments, coloured aggregates, limestone, mica, other crushed rocks, coloured glass) and with its outer structure (washing out the surface, surface scrubbing with a scraper, bush hammer or other masonry tools). [6–8].

From a heritage care point of view, the issue of research into and repairs of hard renders is one of great urgency. Thanks to the marked durability of cement material, a large number of early 20th-century renders have been preserved. However, these facades are nearing the end of their lifespan and require urgent renovation. The structure and colouring of widely applied substitutes today do not correspond to the original material. In order to achieve an optical match of a façade surface, which combines the original material with modern additions, an overall paint layer is usually applied. However, in the case of cement renders this is to the detriment of the original aesthetic concept as the colouring of the façade was achieved through the use of selected aggregates or pigments colouring the plaster throughout its entire mass.

Whereas the issue of other building construction binders has been addressed in this country and abroad, there is a lack of literature dealing with the issue of historical cement-based renders. Much space has been dedicated to lime-based technologies[1] through a whole range of seminars and conferences[2]. Research into Roman cement and its applications has been published in detail within the framework of the outcomes of the international ROCARE project[3]. Some information concerning hard renders can be found in old literature, in masonry textbooks and brochures of the period, which are, however, focused on the issue from a limited point of view of recommendations for craftsmen. Compared with the situation in the Czech Republic, considerably more research is carried out in Belgium, concentrating on hard renders and issues concerning the restoration of architecture of the first half of the 20th century. Some of that research, cited in this article, can serve as comparative material for the study presented here.

## 2. Selected examples of Modern architecture

All samples were taken on the territory of the city of Brno. The reasons were logistical, but Brno was also extremely suitable thanks to the large representation of modern architecture. The buildings selected for the basis of this study, and from which samples were taken for the analysis of the historical renders, are listed in table 1. This selection reflects several aspects; the aim was to cover as many different types of rendering as possible, include a cross-section of buildings typical of the period, and take samples of authentic materials, but not of those on renovated buildings, as this

[1] To cite just one of the dozens of local publications: Girsa V. and Michoinová D. 2015 *Tradiční vápenné omítky: poznání a praktické postupy* (Praha: IC ČKAIT)

[2] Just one of many seminars: „Vápenné omítky pro památkovou péči“ (Lime renders in heritage care) organized by WTA CZ in cooperation with the Klokner Institute of ČVUT and the Faculty of Civil Engineering VUT in Brno, 9. December 2009.

[3] For details see: www.rocare.eu.

would necessitate the inclusion of facades that had previously undergone overall repairs into the survey. However, it was impossible to achieve this aim in full. During the first two decades of the 20$^{th}$ century, lime-based rendering technologies were the norm and acquiring an authentic sample was simply not possible. Not many buildings were constructed during WWII, and on most of those surviving, the renders had been renewed. This was also apparent when taking samples from the buildings of the 1920s and 1930s – the authentic renders were not always found, even though the relevant documentation[4] suggested that they should have been, and in some other cases there are serious doubts as to their actual existence. In spite of these obstacles, samples were taken from significant buildings, documenting the development of modern architecture and building engineering before WWII.

The list of buildings is presented according to a chronological key dating the construction of the building in order to compare the technologies and material composition of the samples taken. Along with the basic information concerning the construction of the building, the table contains a brief description of the render samples and their basic characteristics.

**Figure 1.** (a) Nemocenská pokladna soukromých úředníků a zřízenců (Treasury for social care of private clerks), Kounicova 299/42, Zahradníkova 299/10, Brno (architect: Jaroslav Syřiště, 1923–24), (b) Městské lázně (Municipal Spa), Zábrdovická 158/13, Brno (architect: Bohuslav Fuchs, 1929–1932).

[4] Full use was made of research, specialist opinions, photographs and further documentation of the National Heritage Institute, Regional Specialist Department in Brno.

**Figure 2.** (a) Nádražní poštovní úřad (Railway Station Post Office), Nádražní 118/7, Brno (architect: Bohuslav Fuchs, 1937–1938), (b) sampling place, the front (west) façade.

**Table 1.** Description of the sampling sites.

| Object | Location (district, street) | Architect | Sampling site/position |
|---|---|---|---|
| **Nemocenská pokladna soukromých úředníků a zřízenců (Treasury for social care of private clerks)** | Kounicova 42/Zahradníkova 10, Brno | (1923-24) Jaroslav Syřiště | **A**-terazzo of the plinth **B**-plain surface finishes |
| **Městské lázně (Municipal Spa)** | Zábrdovická 12, Brno | (1929-32) Bohuslav Fuchs | Plain surface finishes |
| **Nádražní poštovní úřad, západní fasáda (Railway Station Post Office, west facade)** | Nádražní 118/7 | (1937-38) Bohuslav Fuchs | Plain surface finishes |
| **Nádražní poštovní úřad, jižní fasáda (Railway Station Post Office, south facade)** | | | Plain surface finishes |
| **Odborná škola pro ženská povolání Vesna (Secondary School for Female Occupations Vesna)** | Lipová 232/1, Brno | (1929-30) Bohuslav Fuchs, Josef Polášek | *terrazzo* of the plinth |
| **Zemský dům III / činžovní dům (National House III/Apartment block)** | Kounicova 944/1-944/7, Brno | (1924) Karel Náhůnek | *terrazzo* of the plinth |
| **Bytový dům (Apartment block)** | Botanická 825-828/37-45, Brno | (1955-1956) Vítězslav Unzeitig | *terrazzo* of the plinth |

The principal method of material investigation was light microscopy (LM). The samples were impregnated with epoxy resin under vacuum and, subsequently, thin sections of 30 μm thickness were produced from their cross sections. The thin sections were examined in transmitted polarized light using an Olympus BX 51 microscope and scanned using a Canon 9000F Mark II film scanner equipped with polarizing foils. One sample was examined also by powder X-ray diffraction with a Rigaku Smartlab diffractometer with Bragg-Brentano Θ-Θ reflection geometry. Ni-filtered Cu radiation ($\lambda_{K\alpha}$ = 0.15418 nm) was used. The tube was powered at 40 kV and 30 mA. Divergence sit width: 0.5°, Soller slits: 5°

receiving slit: 0.1°. Measured angular range: 5-80°2Θ, step size: 0.02°2θ, time per step: 1 s. The gained data was evaluated with the help of Panalytical High Score 3 Plus and Bruker AXS Topas 4 software. The quantitative phase analysis was done by the Rietveld method. Crystal structure models from the Bruker AXS database were used. The analysis of pigments were performed by scanning electron microscopy and X-ray microanalysis (SEM/EDS) using a Tescan Vega TS 5136 LM instruments powered at 30 kV. The samples were carbon coated.

## 3. Composition of hard renders

**Pokladna soukromých úředníků a zřízenců (Treasury for social care of private clerks), Kounicova 299/42, Zahradníkova 299/10, Brno (Jaroslav Syřiště, 1923-24)**

***Sample A – reddish terrazzo of the plinth sections***

The sample is composed of one layer of reddish render (applied to a light ochre, hard, rough-grained core render on the building) containing dark green and white clusters of filler (figure 3a, b). The binder is of a mixture based on aerial lime and Portland cement. There is a dark red oxide pigment quite homogenously dispersed in the binder, locally forming clusters (figure 3d). The pigment is most likely a secondary industrial material containing hematite, e.g. mettalurgical dust. The filler is primarily composed of macroscopic white, crushed limestone and dark green serpentinite (figure 4c) mixed to a volume ratio of 3:2. Alongside these rock clusters there are sporadic traces of quartz, metaquartzite and minute fragments of metabasites. The filler is granulometrically well sorted with two separate particle sizes: 5.0–3.0 mm and 1.0–0.1 mm. With the help of image analysis, the filler to binder volume ratio was determined to be 2:1.

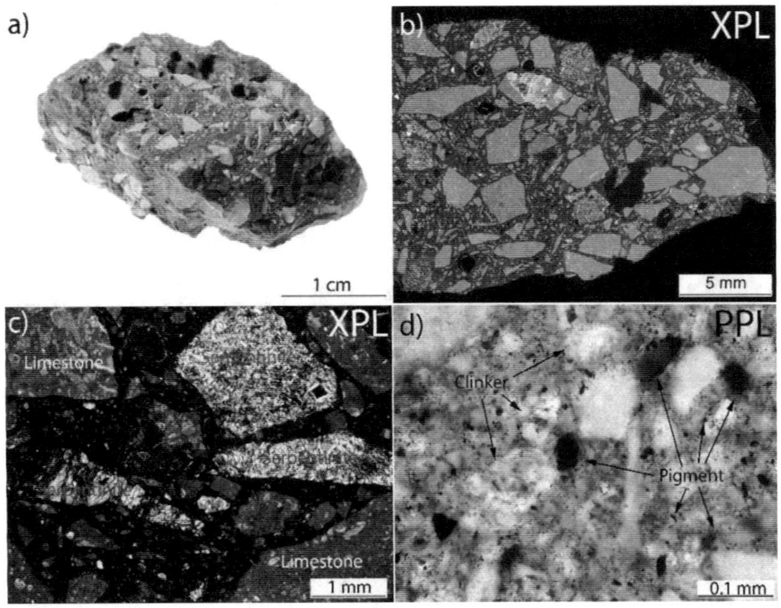


**Figure 3.** Treasury for social care of private clerks, Kounicova 299/42, Zahradníkova 299/10a, *terazzo* of the plinth sections: a) Macro photographs of the fractured sample, b) thin-section photomicrographs of the sample, XPL, c) the detail of the clusters of the aggregate, thin-section photomicrograph, XPL, d) the fragments of the Portland cement clinker particles and hematite pigment, PPL.

***Sample B – plain surface finishes***

The strata of the renders are made up of a light grey-ochre core layer (1) and a greenish surface renders (2) (figure 4a, b). The binder of the core layer is of Portland cement most probably with an admixture of lime whose presence can be surmised from the optical character of the matrix, which displays an increased porosity and low presence of cement clinkers. The top layer only contains a lime binder (the content of unfired clusters of lime is significant). The core mortar was made by mixing the aggregate and binder to the approximate ratio of 3:1, with the same "recipe" used in the top render layers containing lime binder. The aggregate is almost exclusively made up of clusters of serpentinite (figure 4c). Only on the base of the sample are there quartz and metaquartzite clusters, which are not however, part of a separate layer. The granulometric sorting of the aggregate is relatively poor. The size variation in the fragments of filler is <0.1–4.0 mm with relatively uniform particle size distribution. The colour shade of the top render layer was achieved through the addition of a green pigment, which is dispersed throughout the binder (figure 4d). The sporadic presence of glauconite was optically identified within the pigment. The composition of the pigment was analysed by means of EDS which confirms that majority of the green pigment was composed of elements lighter than sodium, which in turn points to the use of an organic dye. The size of the largest pigment grains is up to 0.1 mm, with the majority of the grains however, significantly smaller. The pigment has even penetrated into some of the filler fragments where it has created a relatively extensive light-green tinted area.

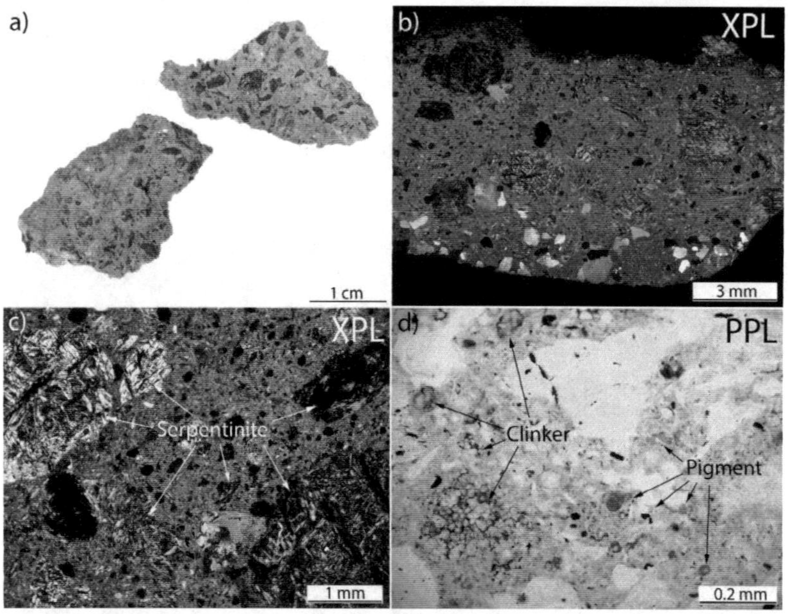


**Figure 4.** Treasury for social care of private clerks, Kounicova 299/42, Zahradníkova 299/10a, plain surface finishes: a) Macro photographs of the fractured sample, b) thin-section photomicrographs of the sample, XPL, c) the detail of the clusters of aggregate – serpentinite, d) the detail of the Portland cement clinker particles and green pigment, PPL.

**Městské lázně (Municipal Spa), Zábrdovická 158/13, Brno (Bohuslav Fuchs, 1929-1932)**
***Plain surface finishes***

The sample consists of a grey sample of render with a surface features macroscopically visible, characteristic protruding aggregate fragments (figure 5a). The strata of the sample are composed of four layers (figure 5b) including the medium-grained glauconitic sandstone as a ground material (layer 0) and 3-layered render with the strata: under-layer (1), core render (2) and stucco render (3). Whilst the under-layer is a lime-cement one, the other render layers feature binders predominantly with containing Portland cement components (figure 5c). The estimated volumetric binder to aggregate ratio of the under-layer is approximately 3:1, with the core and stucco ratios both reaching an identical 1:1. The filler of the core layer (2) is a rich mixture of rock fragments (quartzite, metaquartzite, metamorphic slate, granitoid rock, aplite, sandstone, etc.) and minerals (quartz, alkali feldspar, plagioclase, muscovite, biotite) with an admixture of (probably) blast furnace slag.

The structure and coloured mosaic of the surface layer is made up of a mixture of macroscopic white crushed limestone and black serpentinite mixed to a volume ratio of 1:3, sorted to a grain size of 1–3 mm (figure 5d).

**Nádražní poštovní úřad (Railway Station Post Office), Nádražní 118/7, Brno (Bohuslav Fuchs, 1937-1938)**

***Plain surface finishes (west façade)***

The sample of the render features a distinctive surface structure as a result of large, projecting fragments of the aggregate, visible even under the covering layer of whitewash (figure 6a). The sample taken on site is composed of three layers in total (figure 6b): the core mortar with a fine filler (1) which base has not been captured in the sample, an upper layer 10–14 mm thick with large fragments of filler projecting from the layer of the sample (2). On top, there is a thin, discontinuous, extremely porous, surface layer (3) with thickness varying from 0.4–0.7 mm, which is probably a secondary modification. The binder of layer 1 is Portland-cement compared to layers 2 and 3 which contain lime as a binder. The binder/aggregate ratio of layer 1 was established to be roughly 2:3, with the filler made up of a mixture of crushed limestone of 0.2–2.5 mm particles and finer sand with the gain size distribution of some 0.05–1.5 mm. The aggregate composed of fragments of minerals and rocks - quartz, plagioclase, mica, garnet gneiss, metamorphic slate, flints, granitoid rock and others.

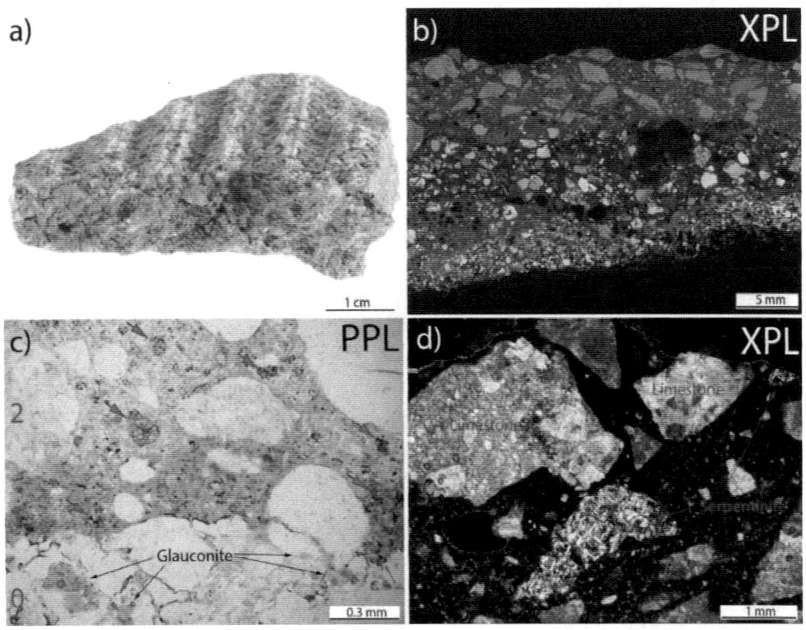


**Figure 5.** Municipal Spa, Zábrdovická 158/13, plain surface finishes: a) Macro photographs of the fractured sample, b) thin-section photomicrographs of the sample, XPL, c) the detail of the stratigraphy (0-glauconitic sandstone, 1-lime-cement render and 2-cement finish (with the details of the Portland cement clinker particles), PPL, d) the detail of the third layer containing crushed limestone and black serpentinite, XPL.

The limestone to sand ratio was estimated to be approximately 1:1. In comparison, the binder to aggregate ratio of layer two is around. The aggregate contains two distinctive grain size regions – a finer one with a prevailing particle size in the region of some 0.05–0.25 mm, composed of marble dust with, sporadically, fragments of sand (figure 6c) and rough clusters of aggregate with particle size 4–16 mm containing sub-rounded to sub-angular fragments of metaquartzite to quartz cathaclasites having low sphericity. Finally, layer 3 was formulated by mixing binder and aggregate to the ratio of 3:1 in volume parts. The porosity of layer 3 is over 30%, probably a result of the addition of an unidentified non-foaming substance (figure 6d). The aggregate is composed of very fine-grained particles of crushed marble dust with particle size distribution 0.02–0.04 mm.

***Plain surface finishes (south façade)***
The sample taken from external wall render consists of one single layer of highly porous render with macroscopic signs of fragments and open porosity onto which a coat of white paint has been applied (figure 7a). Most of the volume of the sample is composed of a layer with a cement-lime binder with a predominance of Portland cement (1). On the surface there is a thin (0.1–0.7), probably secondary, discontinuous paint layer (2) with a lime as a binder and with porosity of over 50% (figure 7b, d). Both the microstructure and composition of the aggregate of the layer (2) are identical to those of layer 3 of the west façade of the building. The render was formulated in volumetric binder to aggregate ratio being about 1:3. The aggregate underwent well sorting; two particle size region can be identified in the filler one being less than 1 mm and the second 3–3.8 mm. The mineralogical composition of the aggregate shows a prevailing mixture of quartz and crushed marble mixed in rough volume ratio of 1:1. There were also sporadic traces of minor inert components identified in the sample (probably blast-furnace slag) and opaque minerals (figure 7c).

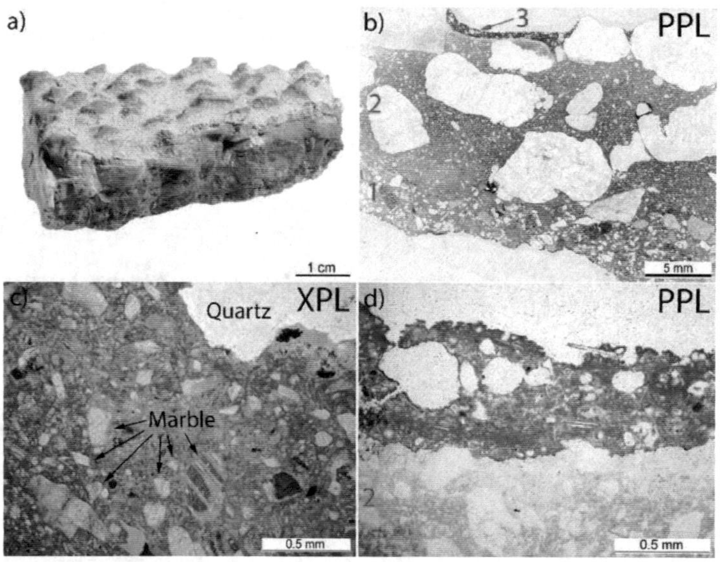


**Figure 6.** Railway Station Post Office, Nádražní 118/7, plain surface finishes (west façade): a) Macro photographs of the fractured sample, b) thin-section photomicrographs of the sample, PPL, c) thin-section photomicrographs, the detail of the marble aggregate, XPL d) the interface between the ground layer (1) and highly porous surface paint layer (2), PPL.

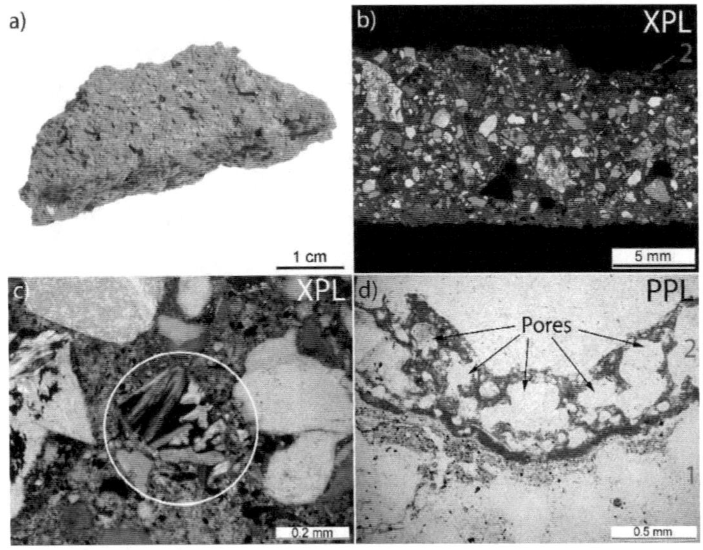


**Figure 7.** Railway Station Post Office, Nádražní 118/7, plain surface finish (south façade): a) Macro photographs of the fractured sample, b) thin-section photomicrographs of the sample, XPL, c) the detail of the slag particle, XPL, d) the interface between the ground layer (1) and highly porous paint layer (2), PPL.

**Secondary School for Female Occupations Vesna, Lipová 232/1, Brno (Bohuslav Fuchs, Josef Polášek, 1929-1930)**

***Terrazzo of plinth***

The render strata consist of ochre core render followed by the upper layer made up of a macroscopically visible red colour and featuring an inhomogenous texture (figure 8a). Under close microscopic inspection, the render strata is actually composed of three layers (figure 8b): two macroscopically visible, light yellow layers of core plaster (1) and its pigmented surface layers (2) and the upper render with red pigment (figure 8c). The base of layer 1 is not captured in the sample, the thickness of layer 2 varied from 1-11 mm. The thickness of the red surface layer (3) is from 9–12 mm. The binder of all three layers is Portland cement, in layer 1 probably with an addition of lime. The binder to aggregate ratio in layers 1 and 2 is identical at approximately 2:3, with the ratio in layer 3 at approximately 1:3. The filler in layers 1 and 2 are identical and is approximately 2:3 and the binder/aggregate at 1:3 in layer 3. The aggregate of layers 1 and 2 are identically composed of sand with a particle size distribution 0.05–3 mm. The sand grains are composed of varied mineral rock fragments (quartz, plagioclase, sporadically muscovite, biotite, glauconite, garnet, sandstone, greywacke, siltstone, metaquartzite, cataclasite, granitoid rock, and vestiges of limestone). The aggregate of top layer is composed of marble dust and sand mixed to a volume ratio of approximately 2:1. The aggregate of the layer is rather medium sorted with grain size distribution 0.1–1.3 mm. The red pigment of layer 3 is hematite, according to the morphology probably a calcined ochre pigment. Within all layers, some of the pores are partially or completely filled with ettringite, probably due to the alteration of the cement binder through exposure to the aggressive atmospheric environment (figure 8d).

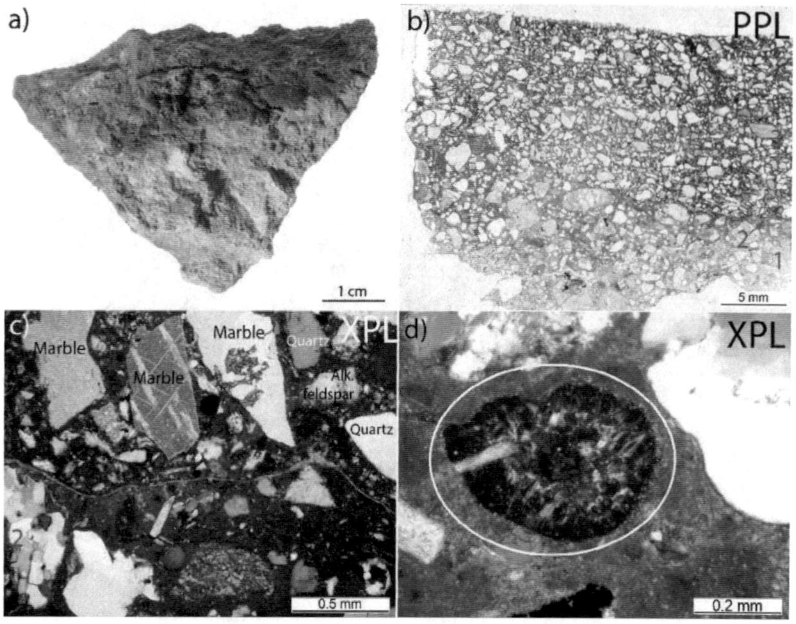


**Figure 8. Secondary School for Female Occupations Vesna, Lipová 232/1**, *terrazzo* of the plinth: a) Macro photographs of the fractured sample, b) thin-section photomicrographs of the sample, PPL, c) the interface between layer the layer the ground layer (2) and red finish (3) (containing crushed marble and sand as an aggregate), d) the detail of the pore filled with ettringite.

**National House III / Apartment block, Kounicova 944/1-944/7, Brno (Karel Náhůnek, 1924)**
***Terrazzo of the plinth***
The macroscopic composition of the sample is of two layers: a grey core render (1) and a surface layer (2) with a greyish-yellow binder, white clusters of filler and a grooved texture (figure 9a). Only layer 2 has been studied microscopically. The binder of the *terrazzo* layer is Portland-cement based, with the binder matrix containing abundant fragments of Portland clinkers of up to 0.3 mm in size. The binder to filler volume ratio is approximately 1:3. Most of the filler fragments fall into two particle size regions, < 1 mm and 2.5–6 mm, both particle sizes were mixed in an approximately 2:3 volume ratio.

**Figure 9.** National House III/Apartment block, Kounicova 944/1-944/7, *terrazzo* of the plinth: a) Macro photographs of the fractured sample, b) thin-section photomicrographs of the surface layer (2), XPL.

The prevailing aggregate component is crushed limestone with the rare occurrence of quartz (figure 9b). The sample contains an ochre pigment composed of iron oxide-hydroxide, most likely of natural origin. The argument for the natural origin is supported by the fact that some clusters of iron oxide-hydroxides occur in the original position – surrounding the minute grains of quartz.

**Apartment block, Botanická 825-828/37-45, Brno (Vítězslav Unzeitig, 1955-1956)**
***Terrazzo of the plinth***
The macroscopically compact sample features a red binder with black and white clusters of aggregate and a rough surface texture (figure 10a). The sample is composed of two layers: the core mortar (1) and the top, approximately 10–13 mm thick (figure 10b) layer with red pigment (2). The binder to aggregate volume ratio in the core mortar is 1:3, with a lower ratio in the top layer (approximately 1:2). The binder of the core mortar is Portland-cement based with poorly sorted aggregate with the majority of the fragments smaller than 2 mm. The composition of the aggregate is of a mixture of sand composed of a rich mixture of mineral and rock fragments and a smaller portion of crushed limestone. On the basis of the PLM and XRD results, in the top layer is Portland-cement based with an addition of gypsum. Using XRD, the amount of gypsum in the whole sample (including the filler) was found to be 7.9 wt. %, therefore, it is unlikely that it would originate from hydrated cement. As a result of the presence of gypsum, ettringite has formed within the sample. The filler is composed of macroscopic white crushed limestone and dark crushed serpentine. The limestone to serpentinite volume ratio is approximately 2:1. Fragments of other rocks and minerals, such as quartz, plagioclases, glauconite and occasional fragments of mica schist, appear only sporadically. The particle size distribution of the aggregate falls predominantly into two regions with most falling between 2–3 mm and less in particle size. The results of the XRD analysis of layer 2 are in listed in table 2.

**Figure 10.** Apartment block, Botanická 825-828/37-45, *terrazzo* of the plinth: a) Macro photographs of the fractured sample, b) thin-section photomicrographs of the sample, XPL.

**Table 2.** Mineral composition of red *terazzo* layer (2) determined by means of XRD in wt.%.

| Aggregate | | Binder +aggregate | | Pigment | Aggregate | | | | | | |
|---|---|---|---|---|---|---|---|---|---|---|---|
| | | | | | | Serpenitite | | | | | |
| Alite | Ettringite | Gypsum | Calcite | Hematite | Quartz | Lizardite | Antigorite | Olivine | Pyroxene | Amfibole | Chlorite |
| 0.4 | 0.9 | 7.9 | 70.0 | 0.4 | 1.4 | 9.9 | 0.6 | 1.4 | 1.7 | 2.2 | 3.0 |

## 4. Conclusions

The study discusses the detailed composition of several samples of renders taken from the façades of buildings constructed in the first half of the 20th century in Brno. The study is the first comprehensive study in this country concentrating on the issue of material composition of renders in modern architecture. The authors would like to continuously expand the topic through further research in order to broaden the spectrum of knowledge concerning the technologies and craft processes involved in preparing the renders on buildings of modern architecture. Although this study is based on a limited number of samples, several interesting assertions can be made, especially as regards the composition of the renders, which are only partially in agreement with those presented in specialist works of reference. Such works of a historic nature tend to address the issue from the point of view of the necessary crafts and skills, whilst overlooking the wide range and changeability of the material recipes. For several reasons therefore, the information acquired is crucial for gaining an insight into the issue at the current time. A subsequent benefit of the analyses of the samples is that they serve as a valuable record contributing to an understanding of the degree of skill of the craft, the materials used and the recipes at the time in question.

The components found in the samples of render can, for example, point to local sources of material that are no longer exploited or indicate the use of, from today's point of view, curious waste material (e.g. slag, pigments, and glass). And lastly, a detailed analysis of the render is crucial in providing a base for suggesting adequate supplementary materials for the restoration of architecturally important buildings of the first half of the 20th century, even in cases where the original material has not been preserved. In extreme restoration situations, the acquisition of knowledge of the techniques and the recipes for materials can be of assistance when devising replicas or allusions for architecture of the first half of the 20th century.

## Acknowledgements

This study is carried out within the frame of the Project of Ministry of Industry and Trade Czech Republic, Programme: TRIO, id. FV30150: *'New Composite Materials for Restoration of buildings of 20th century Architecture'*. The authors would also like to acknowledge Zoja Matulikova from the National Heritage Institute for acquiring several samples of hard render for the purposes of this study.

## References

[1] Kozlowski R, Hughes D and Weber J 2010 Roman cements – key materials of the built heritage of the nineteenth century *Dan M.B., Přikryl R., Török Á (eds) Materials, Technologies and Practice in Historic Heritage Structures* (Springer, Dordrecht) pp 259–257

[2] Weber J, Bayer K and Pintér F 2012 19th century 'Novel' Building Materials: Examples of Various Historic Mortars under the Microscope *Historic Mortars. Characterisation, Assessment and Repair* Valek J, Hughes J J and Groot Caspar J W P (Eds.) Springer pp 89–103

[3] Fiedler F 1957 Teraco a teracové výrobky: Určeno pracujícím ve výrobě, distribuci a použití teraca (Praha: SNTL) p 148

[4] Hogson F T 1906 *Concretes, cements, Mortars, Plaster &Stucco* (Chicago: F. J. Drake) pp 35–45

[5] Dekeyser L, Verdonck A and De Clerq H 2017 Belgian craftsmanship in the interwar period: formulas and application techniques of decorative cement-based wall finishes *International Journal of Architectural Heritage* **11** (5) pp 621–635

[6] Hlaváček B 1914 Příručka zednického řemesla. K poučení a odbornému vzdělání učňů zednických Společenstvo stavitelů pro obvod obchodní a živnostenské komory v Praze pp 23–27, 96–99

[7] Dekeyser L, Verdonck A and De Clerq H 2013 Cimorné cement render with opalescent glass granules: a decorative façade finish development by innovative craftsmanship in the interwar

period *Journal of Architectural Conservation* **19** (2) pp 86–102
[8] Fiedler F 1957 Teraco a teracové výrobky: Určeno pracujícím ve výrobě, distribuci a použití teraca (Praha: SNTL) p 148
[9] Govaerts Y, Verdonck A, de Bouw M and Meulebroeck W 2014 The introduction of ready-mix rendering mortars for stone imitation in Belgium [1920-1940] *Proceeding of the First Conference of the Construction History Society* (Cambridge: 11-12th April 2014) pp 181–190
[10] Knor J 2015 *Obnova omítek meziválečné architektury* (Brno: Národní památkový ústav) pp 11–16
[11] Valdhansová L et al. 2017 Brněnský architektonický manuál: průvodce architekturou 1918–1945
[12] Kroupa P et al. 2017 *Vila Stiassni: 90 pohnutých let* (Brno: Národní památkový ústav) pp 5–11

# Characterization of fillers made of natural stones as a cement substitute

**L Musil[1], T Cibulka[1], J Vodicka[1] and R Chylik[1]**

[1] Faculty of Civil Engineering, Czech Technical University in Prague, Thakurova 7, 166 29 Praha 6, Czech Republic

E-mail: lubos.musil@fsv.cvut.cz

**Abstract**. The article deals with the classification of fillers according to the type and method of rock processing and their use as a partial cement replacement in the concrete mixture. The aim is to better classify fillers by their origin, maximise their usability and design a concrete mixture with effective use of filler as a cement replacement.
The article compares the fillers that originated from natural stone. The research is focused primarily on the granite filler, as it is not currently explored in detail. As the results of existing research differ, tests of the filler and its effect on the compressive strength of cement composites were performed, for better characterization of the effect of filler pozzolanic activity. The results of the tests are compared with each other, and the probable causes of differences in the individual research results are described.

## 1. Introduction

Nowadays, many types of filler are being researched. The primary reason is the saving cement in composites which reduces the price and energy demands of concrete production. Many of the fillers are also secondary or waste materials. Their processing eliminates the problem of landfill. In general, fillers act in cementitious composites as inert addition. They are considered to be chemically inactive and are added to the concrete mixture primarily to improve processability, increase the proportion of fine cement matrix, change the colour of the composite, and the like [1]. Research shows that fillers can have a positive impact on other properties of hardened composites – for example, permeability, porosity, and strength characteristics. The final composite characteristics are highly dependent on the filler used. There exist many research studies on seemingly the same fillers that disprove each other. Therefore, the detailed characteristics of the filler is an essential part of such research (composite design). In addition to the essential characteristics required by EN 12620 + A1 such as bulk density, grain size, chloride, sulphate content, filler purity [2], detailed chemical analysis or mineralogical composition, grain shape, specific surface area and the like are needed, since all these characteristics are dependent on the type of filler.

The primary classification of fillers depends on the origin of the material. These can be human-made materials (brick, ceramic, concrete recycled materials [3]) or natural raw materials. This article describes fillers formed from natural raw materials – rocks. Fillers can be divided according to the types of rocks from which they originate, which has a significant effect on their characteristics. Fillers can also be divided into several groups, according to the methods of their formation. Initial formation where the filler is the desired product and is produced by micronising. Secondary, when the filler is produced as a waste (secondary) raw material. The secondary filler can be further subdivided, for example, according to the method of rock processing. The rock can be crushed into aggregates which creates dust.

The rock can be processed and treated underwater to form a decorative stone. This process produces sludge which must be dried. Those examples show that there are a large number of filler types. Each has specific characteristics and hence different influence in cement composites.

Most of the filler arises probably when the rock is worked on decorative stone. According to some research, up to 58% of the extracted rock is wasted, of which 18% is waste (sludge) resulting from polishing [4]. This value is very individual as it depends on the method of rock extraction and the final product. At the same time, however, it shows how much total waste can be generated. Other research shows that the amount of sludge produced by cutting blocks ranges from 20 to 30% of the rock volume [5]. The documents suggest that the extraction and especially the modifying of decorative stone blocks generate large quantities of fine dust. The problem with landfilling of these wastes is related to it.

In the article, firstly, fillers from fundamental rocks are compared. As many types of fillers may be formed, the article deals in more detail only with granite sludge resulting from wet stone processing (cutting, grinding, polishing). This material was chosen because it has not yet been examined in more detail as other fillers such as marble or limestone. The researches of the effect of the granite dust on cement composite that has been carried out are often contradictory. The article, therefore, draws attention to characteristics and the more detailed classification of fillers, which has a significant effect on the cement composite.

## 2. Materials and methods

Marble and granite dust was included among the studied materials as they are the most common rocks used as a decorative stone. Furthermore, limestone flour, which is most often used in self-compacting concretes, was compared. Other fillers were chosen concerning aggregate mining in the Czech Republic – gneiss, amphibolite. In this study, the characterization of granite and amphibolite dust have been performed, and the results were compared with other studies focused on marble, gneiss and limestone dust [4, 7–17].

### *2.1. Grain curve*

The essential physical property of each aggregate, the grain size curve is usually measured on sifting systems. According to European standards, the grain filler test should be carried out as with aggregate by screening with an air jet (airflow) (ČSN EN 933-10) [2]. The ideal solution for the determination of the granularity curve, as was done here for granite dust, is the laser diffraction method, which gives more accurate and detailed values. Also, essential statistical data such as modus, median, arithmetic grain diameter or standard deviation can be determined.

### *2.2. Specific surface*

At first, a specific gravity test was performed using a pycnometer. Depending on this value, other material properties such as specific surface area can be calculated. A separate test of the specific surface was carried out by permeable method using Blain apparatus.

### *2.3. Shape of grains*

The shape of the grains is a significant feature for use in cement composites. This test is not always carried out, but the shape of the filler grains has a significant effect on the fresh mixture (rheology, processing) as well as on the hardened composite. The tests were carried out using software on electron microscope images. In the paper, granite and amphibolite dust were examined using the software. The tests were performed on more than ten samples, and the number of particles tested was always given. Circle equivalent diameter, Aspect ratio and Circularity were measured. The circularity can range from 0 to 1, with a perfect circular shape having a value of 1 and elongated shapes having values close to 0.

### *2.4. Chemical analysis*
The EDS (Energy Dispersive Spectroscopy) analysis was performed on the samples. Multiple samples from each rock were tested, and a more significant amount of grain (area) had to be measured. In the case of smaller focus and larger grain, the analysis was biased. The measured values were compared with those of other researches from India, China, and others. This method is comprehensive, so there is a wide range of results suitable for comparison.

### *2.5. Measurement of pozzolanic activity*
The test of pozzolanic activity may be direct and indirect. Both were used in this research. This measurement is customary for chemically active admixtures in concrete, such as fly ash, metakaolin or micro silica. As the filler is only considered as an inert material, this test (direct pozzolanic activity in particular) is not typically performed. For comparison of individual fillers, it is advisable to include the exam in the research. It shows the chemical activity of fillers, possibilities of use and suitability for use in cement composites.

*2.5.1. Direct pozzolanic activity.* There are several types of tests. A modified Chapelle test was used for this purpose. The method is based on the pozzolanic ability to combine with CaO. The Chapelle test result is expressed as the amount of $Ca(OH)_2$ in mg bound to 1 g of pozzolan. A material is considered pozzolanic if the test result is at least 650 mg $Ca(OH)_2$ $g^{-1}$ [3, 6]. Tests were carried out on dust from the quarries in the Czech Republic (amphibolite), granite sludge and its heat treatment at 800°C with a duration of 2 hours. For marble and limestone, this test makes no sense since they do not contain $SiO_2$.

*2.5.2. Indirect pozzolanic activity.* From these methods, the modified Strength Activity Index (SAI) for fly ash and micro silica, the activity index, was selected. The tests are based on the standards ČSN EN 450-1 (722064) Fly ash for concrete. Part 1 and EN 13263-1 + A1 (722095) Silica fume for concrete. Part 1. According to standards, three mortar mixtures with standard quartz sand have been designed for cement strength tests. In the first case, a reference sample with only standard sand and CEM I 42.5 cement was designed. In the second case, a mixture was designed according to the standard for the specified efficiency index for fly ash, and the cement replacement weight was 25% granite dust. The third mixture was designed according to the silica fume standard, and the investigated filler replaced 10% cement weight. Strength tests were carried out after 3 and 28 days on prisms 40/40/160 mm and cubes 40/40/40 mm.

## 3. Results
The tests we conducted were partially compared with results from other studies. Among the first characteristics is the grain curve, respectively, the maximum grain of the filler. Some literature reports only the maximum grain (sieve passage). These values are most often given as 0.125 mm; 0.100 mm; 0.075 mm; 0.063 mm; 0.060 mm. Others report the average aggregate grain. The most suitable expression is through a grain curve or particle size distribution which is expressed numerically (e.g. D10, D25, D50, D75, D90). In particular, the D10, D50 and D90 values are published in the papers.

**Table 1.** A particle size distribution in μm.

| filler | citation | D10 | D25 | D50 | D75 | D90 |
|---|---|---|---|---|---|---|
| granite | | 1.901 | 5.002 | 13.202 | 28.83 | 51.946 |
| granite | [7] | | | 18.64 | | |
| granite | [8] | | | 11.40 | | 50.000 |
| granite | [9] | 1.550 | | 13.050 | | 55.460 |
| granite[a] | [9] | 0.840 | | 3.480 | | 13.340 |
| granite | [10] | 3.700 | | 21.400 | | 67.000 |
| gneiss | [11] | 3.461 | | 22.825 | | 86.668 |
| gneiss | [11] | 2.864 | | 18.902 | | 66.811 |
| ampfibolite[b] | [12] | 5.661 | | 54.445 | | 204.691 |
| limestone[b] | [12] | 3.435 | | 43.247 | | 147.624 |
| marble[b] | [13] | 2.300 | | 33.000 | | 350.000 |

[a] filler was produced by micronising
[b] values were subtracted from the graph

For example, if only the maximum aggregate grain is given, it is advisable to supplement the results with another characteristic. In this case, a specific surface can be used. It can also be compared with a cement value, which is approximately known or recorded in the technical datasheet. At the same time, this comparison suggests the suitability of using a filler. According to researches, the specific surface area of fillers is in the range of 300–450 $m^2kg^{-1}$. There are also smaller values of about 240 $m^2kg^{-1}$ [4, 8, 14], especially for fillers of marble and limestone. According to our research, the specific surface of granite dust is 437 $m^2kg^{-1}$.

Another characteristic is the grain shape. According to the SEM figures in the available literature [15], the granite, marble, limestone and gneiss filler grains are similar, which is also evidenced by the aspect ratio from the article [15]. Granite dust and amphibolite were compared in our research (table 2). A significant difference can be seen from the SEM figures (figure 1). As shown in the figure, the vast majority of amphibolite grains are acicular. Poor grain quality is confirmed by a roundness of 0.527, which is lower than that of granite dust.

**Figure 1.** SEM granite dust (left), amphibolite dust (right).

**Table 2.** The grain shape.

| Rock type | Amount of particles tested (pcs) | Circle equivalent diameter (μm) | Aspect ratio (-) | Circularity (-) |
|---|---|---|---|---|
| Granite | 73 | 28.0 | 0.623 | 0.641 |
| Amphybolite | 106 | 23.3 | 0.533 | 0.527 |

IOP Conf. Series: Materials Science and Engineering 1039 (2021) 012007 doi:10.1088/1757-899X/1039/1/012007

The chemical composition of the filler has a significant influence on its reaction in the cement composite. In table 3, it can be seen that a significant difference in chemical composition is not only among the filler types but also the same filler from a different quarry. The difference can be observed even in row 1, where all samples are from the same quarry.

**Table 3.** The chemical composition of the filler.

| filler | $SiO_2$ | $Al_2O_3$ | $Fe_2O_3$ | $Na_2O$ | $K_2O$ | CaO | MgO | LOI |
|---|---|---|---|---|---|---|---|---|
| Granite[a] | 48.6–62.9 | 14.1–14.6 | 6.3–8.0 | 3.0–4.2 | 8.0–9.8 | 2.6–4.5 | 3.5–5.9 | |
| Granite[b] [4, 16] | 53.2–94.2 | 1.3–14.1 | 0.4–12.3 | 1.2–3.6 | 0.1–0.2 | 1.0–9.1 | 1.7–8.3 | 0.3–5.0 |
| Granite[c] | ( ~ 60) | ( ~ 14.0) | - | ( ~ 3) | - | - | ( ~ 2) | - |
| Amphibolite | 44.82 | 14.51 | 13.46 | 0 | 0.49 | 6.38 | 20.33 | |
| Gneiss [15] | 70.13 | 15.95 | 0.9 | 0 | 6.22 | 1.12 | 0.69 | 2.77 |
| Marble[b] [4, 13, 16] | 0.2–14.8 | 0.1–21.9 | 0.1–36.8 | 0.0–9.3 | 0.1–0.2 | 6.0–83.2 | 0.4–9.3 | 2.5–46.0 |
| Marble[c] | ( < 1.5) | (<1.0) | (<0.5) | (<1.0) | (<0.1) | ( ~ 50.0) | ( <0.7) | ( ~ 43.0) |
| Limestone[b] [4, 16, 17] | 0.0–3.3 | 0.3–0.8 | 0.0–0.6 | 0 | 0 | 42.6–92.9 | 0.0–9.6 | 1.2–43.7 |
| Limestone[c] | ( ~ 0.10) | ( ~ 0.30) | - | ( < 0.0) | ( < 0.0) | ( ~ 50.0) | ( ~ 1.0) | ( ~ 43.0) |

[a] Chemical composition in one quarry
[b] Chemical composition from multiple quarries
[c] Common (most common) values

The reactivity of the filler is related to its chemical composition. Using the modified Chapelle test, it has been shown (table 4) that although materials contain large quantities of silicon, they are not very reactive and do not meet the standard requirements. Granite dust is more reactive than amphibolite and by modification (heating) its reactivity can be increased by about 50%. Even then, it does not comply with the standard.

**Table 4.** Chapelle test.

| sample | mg $Ca(OH)_2$ $g^{-1}$ |
|---|---|
| Granite | 345–374 |
| Modified granite | 535 |
| Amphibolite | 75–150 |

The last test of the filler was an indirect determination of pozzolanic activity. The flexural strength on the mortar prisms three days after concreting reached the highest values for the first mixture, consisting only of sand and cement. After 28 days of concreting, the highest values of flexural strength were measured for mixture with 10% of granite dust as cement replacement (figure 2).

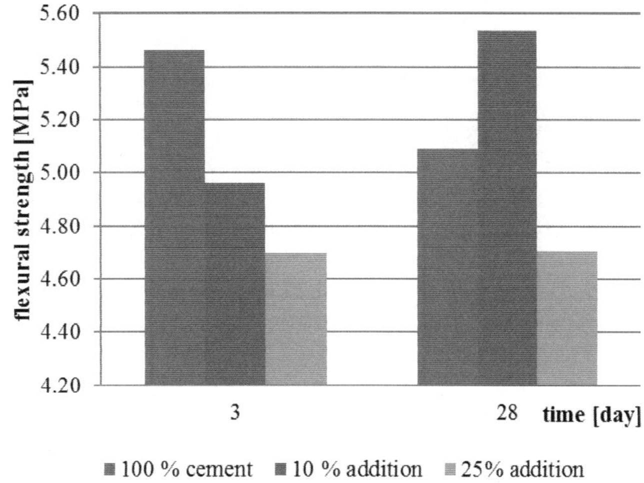


**Figure 2.** The flexural strength on the mortar prisms.

The highest value in compressive strength three days after concreting was achieved with the 25% cement replacement mixture (figure 3). In tests after 28 days, the highest average compressive strength values were measured with a 10% cement replacement mixture. Since direct tests of pozzolanic activity did not meet the standard requirements, higher compressive strength is likely to be due to denser structure caused by the better filling of cement stone with micro-filler.

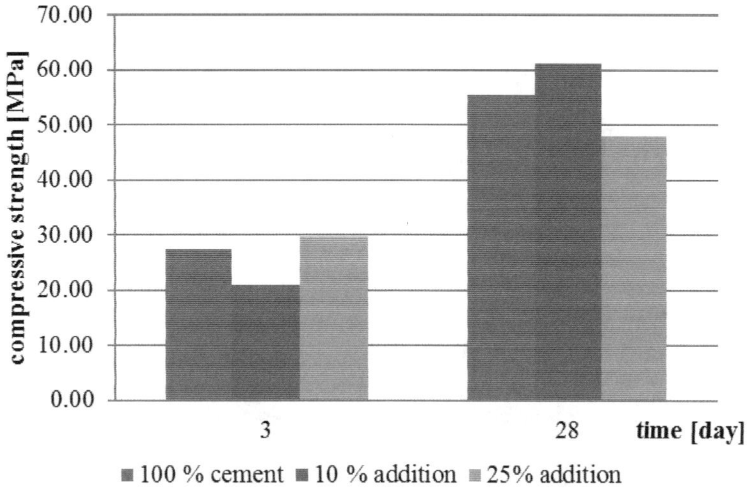


**Figure 3.** Direct pozzolanic activity test – compressive strength.

## 4. Discussion

Grain size and distribution were solved as a fundamental characteristic. All grains are less than 0.125 mm in both research and literature, except amphibolite. Granite sludge had 90% grains smaller than about 55 μm, except for sludge treated with micronising (13.34 μm). The gneiss and amphibolite (crushed) fillers had larger grains. Significantly larger grains of amphibolite are probably due to their shape, where acicular grains are less likely to fall through the sieves. A positive effect of the filler size is given in a paper [9], where the smaller-grain mortar had higher strength than the larger-grain mortar (table 1). Overall, the mortars in this research had lower strengths than the reference mortar, due to an incorrectly calculated water coefficient. A suitable complement to the grain size is the specific surface, which is not often investigated. The results show that the fillers usually have a similar specific surface area as the cement, indicating a proper use of filler in cement composites. According to the literature [4, 8, 16], marble and limestone fillers may have a smaller surface area (approx. 240 $m^2 \cdot kg^{-1}$) than granite dust. All investigated and compared fillers had a similar aspect ratio except amphibolite, which had a lower circularity than granite dust. For these reasons, amphibolite is not suitable for use in cement composites as it worsens their rheology.

In the chemical composition (table 3) the substantial differences can be seen not only between the kinds of fillers but also within a single filler. This characteristic is influenced by the location of the quarry and the homogeneity of the rock. Even in a single quarry, there can be a change in chemical composition during mining. In table 3, common chemical composition values that occur most often are written in brackets. In granite, gneiss and amphibolite fillers, the predominant component was silicon oxides ($SiO_2$) and aluminium oxides ($Al_2O_3$). Conversely, marble and limestone fillers contained 50% calcium oxide (CaO), and a large part consists of loss on ignition (LOI). The direct pozzolanic activity was measured for granite dust and amphibolite as they contain a large amount of silicon. The results confirmed that there is insufficient activity according to the standard (650 mg $Ca(OH)_2$ $g^{-1}$) and therefore it is not mostly amorphous silicon. Granite dust demonstrates significantly better values than amphibolite, and by heat treatment, the pozzolanic activity can be increased by 50% to 535 mg of $Ca(OH)_2$ $g^{-1}$. Still, it does not meet the standard value. On the contrary, the indirect test of pozzolanic activity was positive according to the standard. The efficiency index after 28 days was 110.26% for a 10% cement replacement. It thus meets the criterion of the standard (ČSN EN 13263-1 + A1) according to which it was tested. Positive results were also achieved by the second applied standards (EN 450-1) when the efficiency index exceeded the required value of 75% by 11.43%. Since the direct pozzolanic test was negative, the better strength characteristics are probably due to the denser structure of the composite. These results are also confirmed by a paper [17], where the strength increases with the bulk density of the composite. According to the comparison of available papers, it is clear that the water coefficient has a significant influence on the resulting characteristics of the composite. In cases where the water coefficient was calculated precisely (water weight/cement weight), the strength characteristics were better with the granite filler. In contrast, when calculating the water coefficient, including the filler weight, the strength characteristics of the granite dust composite were smaller than the reference composite. Even so, the research [16] considered the water coefficient including partly the granite filler in its cement ratio, the strength characteristics of some recipes were still slightly positive.

## 5. Conclusions

Granite dust has approximately the same grain size. Compared to the fillers produced by crushing (gneiss, amphibolite), it has smaller grains. The grain size affects the strength characteristics of the cement composites. It is essential to test the specific surface, which is not often tested. The specific surface of the fillers is similar to that of cement. Limestones and marbles may have a smaller specific surface. Grains of granite, gneiss, marble, limestone filler are more suitable for use in concrete composites than amphibolite filler. Its grains are predominantly elongated and have an acicular shape. Therefore, amphibolite is not suitable due to the processability of the mixture.

The chemical composition varies for both multi-quarry fillers and within a single quarry and has a significant effect on reactions in concrete. Granite, gneiss, and amphibolite fillers can be expected to have a partial pozzolanic reaction as they contain a large amount of $SiO_2$. According to the direct pozzolanic activity test, any filler does not meet the standard values for pozzolan. Granite dust has higher pozzolanic activity than amphibolite. At the same time, granite dust can be partially activated by heat treatment. The indirect method showed a positive effect of granite dust on strength characteristics of cement composites. As the direct test result of pozzolanic activity was negative, the better strength characteristics are due to the denser structure of the cement composite. An essential condition is the correct calculation of the water-cement ratio, which is a common problem in other researches, and the strength of the filler mortars is lower than the reference mortars.

**Acknowledgements**
This work was supported by The Grant Agency of The Czech Technical University in Prague, grant number SGS20/109/OHK1/2T/11.

**References**

[1] Hela R 2005 *Technologie betonu I* (Brno: Brno University of Technology)
[2] ČSN EN 12620+A1:2008 Aggregates for conrete (Czech Standards Institute)
[3] Zaleska M, Pavlikova M, Pavlik Z, Jankovsky O, Pokorny J, Tydlitat V, Svora P and Cerny R 2018 *Constr Build Mater* **160** pp 106–116
[4] Rana A, Kalla P, Verma H K and Mohnot J K 2016 *Journal of Cleaner Production* **135** pp 312–331
[5] Careddu N, Siotto G, Siotto R and Tilocca C 2013 *Resour. Policy* **38** (3) pp 258–265
[6] NF P 18-513:2009 Pozzolanic Addition for Concrete - Metakaolin. Definitions, Specifications and Conformity Criteria. Annex A.
[7] Zhang H, Ji T and Lin X 2019 *Constr Build Mater* **211** pp 688–702
[8] Mashaly A O, Shalaby B N and Rashwan M A 2018 *Constr Build Mater* **169** pp 800–818
[9] Ramos T, Matos A M, Schmidt B, Rio J and Sousa-Coutinho J 2013 *Constr Build Mater* **47** pp 1001–1009
[10] Li H, Huang F, Cheng G, Xie Y, Tan Y, Li L and Yi Z 2016 *Constr Build Mater* **109** pp 41–46
[11] Hlavinková E 2012 Potential replacement of blast furnace slag in blended portland cements *Bachelor thesis* (Brno: University of Technology) thesis supervisor Fridrichová M
[12] Gajdoš J 2015 Self-compacting concrete SCC for monolithic construction *thesis* (Brno: University of Technology) thesis supervisor Hela R
[13] Nežerka V, Hrbek V, Prošek Z, Somr M, Tesárek P and Fládr J 2018 *Journal of Cleaner Production* **195** pp 1081–1090
[14] Sadek D M, El-Attar M M and Ali H A 2016 *Journal of Cleaner Production* **121** pp 19–32
[15] Schankoski R A, Pilar R, De Matos P R, Prudêncio L R and Ferron R D 2019 *Constr Build Mater* **224** pp 659–670
[16] Elyamany H E, Elmoaty A E M A and Mohamed B 2014 *Alexandria Engineering Journal* **53** pp 295–307
[17] Vijayalakshmi M, Sekar A S S and Ganesh Prabhu G 2013 *Constr Build Mater* **46** pp 1–7

# Effects of elevated temperature on the behaviour of concrete beams reinforced with fiber reinforced polymers

**J Prokeš[1], I Rozsypalová[2], F Girgle[2], P Daněk[2] and P Štěpánek[2]**

[1] PREFA KOMPOZITY, a.s., Kulkova 10/4231, Brno, Czech Republic
[2] Brno University of Technology, Faculty of Civil Engineering, Veveří 331/95, Brno, Czech Republic

E-mail: prokes@prefa.cz

**Abstract**. Fiber reinforced polymer (FRP) rebars have increasing popularity in the construction industry all around the world although steel rebars are widely used for reinforcing of the concrete so far. FRP bars, which have higher tensile strength compared to steel rebars with the same nominal diameter under normal conditions, are composed of resin matrix and fibers. In this paper, the load-bearing capacity of FRP reinforced concrete after elevated temperature exposition are present. The results are compared with concrete sample with steel reinforcement. Commercially produced glass FRP (GFRP) and carbon FRP (CFRP) rebars with sand coatings surface treatments were implemented in concrete beam and subjected to four-point bending load. The residual flexural strength of reinforced concrete after heating to 1000°C was obtained and evaluated and compared to results of non-heated elements. The results indicate that fire resistance of FRP reinforced beams can reach at least 60 minutes according the standard ČSN EN 13501-2.

## 1. Introduction

The present building industry is increasingly influenced by the requirements for low energy performance and long-life of buildings. Concrete structures are often operated in areas with high chemical load. Otherwise, a lot of waste products is used as cement component, leading to a decrease in the pH of the concrete. Therefore, increasing attention is paid to the fibre-reinforced polymers (FRP) as a corrosion resistant alternative to the steel rebar.

Great attention is paid to fire and temperature resistance of FRP reinforcement. A relatively large number of authors dealing with this issue indicate that this is a key feature of these materials. A behavior of FRP rebars and concrete bond at elevated temperature are very well presented in paper [1]. The authors carried out pullout tests, and obtained the bond-slip curve at the temperature ranging from 20–350°C. They propose model based on the glass transition temperature Tg and the decomposition temperature of matrix. The bond-slip constitutive relationship model at elevate temperature can be used for the analysis of fire performance of FRP reinforced concrete members at early stage of the fire, but the analysis of other stage of the fire and post-fire are not the subject of this paper.

Similar study of the effect of elevated temperatures on the mechanical properties of FRP bars and the bond behavior between FRP bars and concrete was carried out [2]. Four types of reinforcement bars: basalt FRP (BFRP), CFRP, GFRP, and steel bars of 10 mm diameter were used. The results showed that the FRP bars suffered significant reductions in their mechanical properties upon exposure to high temperatures of up to 450°C. At a critical temperature of 325°C, the FRP bars lost as high as

(55% and 30%) of their tensile strength and elastic modulus, respectively. The percentage reduction in mechanical properties and bond strength was more pronounced in specimens with FRP bars than those with steel bars under elevated temperatures; the percentage reductions in bond strength between concrete and FRP bars reached as high as 81.5% after exposure to 325°C.

The paper [3] study residual tensile strength, modulus of elasticity, and bond strength to concrete of GFRP bars after exposure to elevated temperatures of up to 400°C and subsequent cooling to an ambient temperature. The results showed that the residual strength generally decreases with increasing temperature exposure. However, as much as 83% of the original tensile strength and 27% of the original bond strength was retained after the specimens were heated to 400°C and then cooled to ambient temperature.

The authors of the study [4], took a next step and perform subsequent axial tensile tests and pull-out tests of these materials after exposing them into elevated temperature effects in the range of 23–800°C. Severe effects on the tensile properties of bare steel bars were observed after 600°C, while this critical limit is 300°C for bare GFRP bars. Test results show that bond strength degradation is almost linear for both type of rebars, however, 600°C is also the critical temperature regarding the serious deterioration on concrete.

These papers [1–4] and many others are a very good starting point for our study, when several types of reinforcement are casted in concrete and after heating the element to 1000°C is subsequently tested.

## 2. Experimental analysis

An experiment to determine the effect of elevated temperatures on the residual bearing capacity of concrete beams reinforced by FRP or steel reinforcement was carried out. The experiment was performed on concrete elements of size 100 × 150 × 1800 mm reinforced by GFRP, CFRP or steel rebars.

### *2.1. Materials*

All concrete beams were reinforced in flexure with 1 longitudinal bar at the bottom, four types of main reinforcement were used: GFRP rebars 10 and 14 mm with sand coated and braided surface (beams G10 and G14), CFRP rebars 9 mm sand coated and braided (beams C9) and steel rebars 12 mm of class 10505 (beam S12). For longitudinal reinforcing bars in a beam, the concrete cover was 25 mm. GFRP and CFRP rebars are standard product of company PREFA KOMPOZITY. The same epoxy resin is used for the production of these rebars. Glass transition temperature Tg of the resin is 120°C.

Stirrups made of steel rebars with a diameter of 6 mm were bent into the box-shape and were placed symmetrically at a distance of 150 mm from mid-span and then another 4 stirrups were placed with spacing 80 mm. Stirrups were located symmetrically about the center of the beam, a total of 10 stirrups per beam. Assembling reinforcement (steel $\phi$6 mm) was placed at the top of the beam. Anchoring fixtures were placed into the anchoring area on the longitudinal FRP reinforcement [5].

The samples were made of concrete of strength class C 30/37. The mean compressive strength of the concrete and the modulus of elasticity were 40.95 MPa and 27.9 GPa, respectively.

Six beams were produced for each beam type - three for bending test after exposure to elevated temperatures and three as reference for bending tests of unheated elements. A total of 24 reinforced beams were prepared for the tests.

### *2.2. Fire test*

The samples were exposed to high temperatures in a fire furnace to study the high temperature behaviour of materials. The beams were placed on top of the furnace to provide heating from one (bottom) side. The temperature increase in the furnace up to 1000 C was controlled in accordance with ISO 834 [6], the maximum temperature on the samples was maintained for 2 hours.

During the fire experiment, temperatures were measured using a type-K thermocouple fixed to the reinforcement in the heated mid-span and in the unheated anchorage zone of members.

### *2.3. Mechanical loading*
The beams cooled to ambient temperature were subjected to four-point bending test. The samples were placed on a cylindrical support at a distance of 1000 mm. Beams were loaded by two forces at a distance of 300 mm. During loading, vertical deformations of the structure were measured by means of potentiometric displacement sensors (POT1-5). The deflection was measured in five sections along the beam - in the mid-span, under the loading forces and at a distance of 200 mm from the supports. The test geometry is shown in figure 1. The force F acting as two loads was measured and recorded in the test.

**Figure 1.** Four-point bending test of reinforced concrete beams.

## 3. Results and discussion
The reinforced concrete samples were exposed to high temperatures in the furnace. Temperatures in furnace and at reinforcement can be seen in figure 2. The highest measured temperature at the end of fire test at reinforcement was 570 and 140°C in mid-span and in the anchorage zone, respectively. The time-temperature curves show flat area (slightly above 100°C) between 15–45 minutes. It is caused by evaporation of residual moisture. The temperature 325°C often mentioned in the literature as a critical temperature [1, 2, 7] is achieved in mid-span in most cases between 60 and 90 minutes. Although some degree of irreversible polymer degradation has certainly occurred, the decomposition temperature was not reached. The temperature of reinforcement in anchorage zone (in the ends of concrete beam) reaches temperature Tg = 120°C in approximately 150–180 minutes.

The samples cooled to ambient temperature were subjected to four-point bending test. The mechanical tests records of the selected beam are shown in figure 3 in the form of a load displacement diagram. The deflection measured at mid-span is considered to be a displacement. For all samples, exposed to temperature 1000°C is characteristic that the slope of the load-displacement dependence is smaller, compared to unexposed samples. Also, the samples reinforced with steel show a change in slope, which indicates reduction in steel-concrete bond. It can be caused by irreversible temperature change in concrete. The change of slope in load - displacement diagram of CFRP and GFRP reinforced samples is greater than that of steel and also the maximum load bearing capacity is lower. It should be noted that load - displacement diagram of C9_T20 and G14_T20 respectively C9_T1000 and G14_T1000 are very similar. The usage of GFRP can be comparable to users, but more cost-effective option.

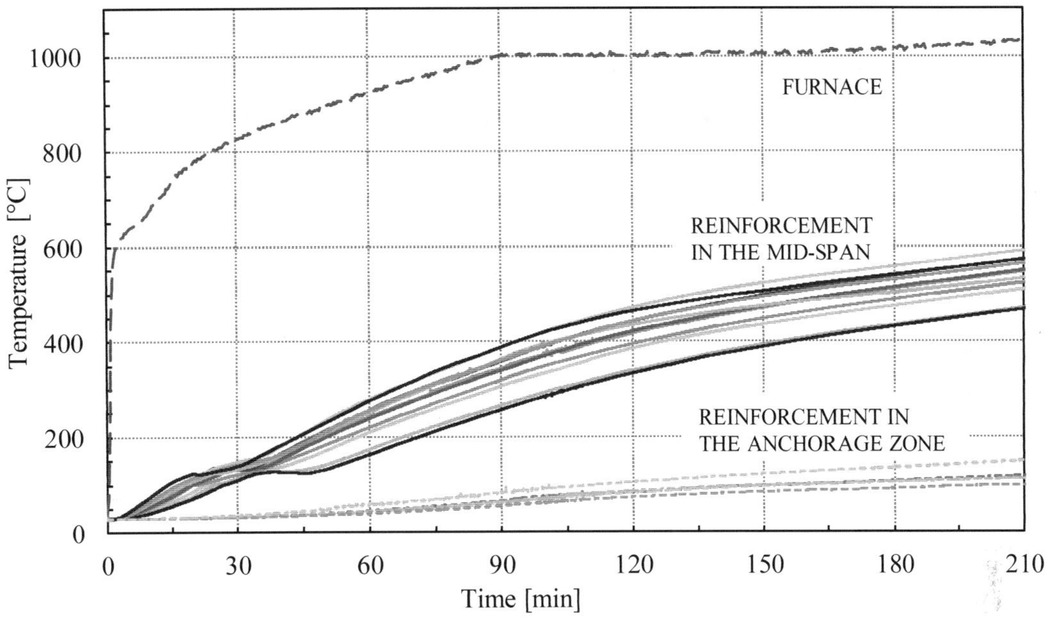


**Figure 2.** Temperatures measured in the furnace and at the reinforcement during fire test.

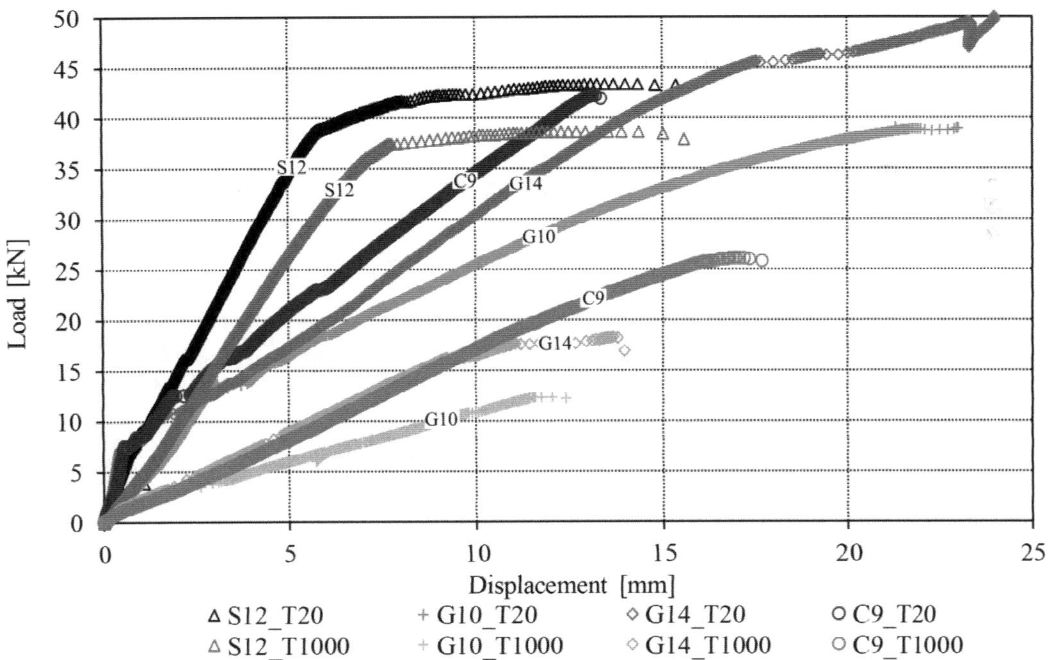


**Figure 3.** Load displacement diagrams in 4-point bending test of concrete beams reinforced by steel, GFRP or CFRP rebars. The symbol "_T20" indicates reference sample which has not been exposed to high temperature. The symbol "_T1000" indicates sample exposed to temperature 1000°C.

The resulting maximum bending capacity determined from the mean of three beams of each group are given in table 1. Also comparison of load-bearing capacity of reinforced beams are shown in table 1 for members exposed and not exposed to high temperatures.

**Table 1.** Flexural bearing capacity of reinforced concrete beams.

| Beam | T (°C) | $F_{max,T}$ (kN) | $F_{max,T}$ / $F_{max,20}$ (%) |
|---|---|---|---|
| S12 | 20 | 44.4 ± 0.9 | 100.0 |
| S12 | 1000 | 40.0 ± 1.4 | 90.2 |
| C9 | 20 | 42.5 ± 3.0 | 100.0 |
| C9 | 1000 | 27.9 ± 2.0 | 65.6 |
| G10 | 20 | 40.5 ± 2.8 | 100.0 |
| G10 | 1000 | 12.6 ± 0.8 | 31.1 |
| G14 | 20 | 41.0 ± 10.7 | 100.0 |
| G14 | 1000 | 18.7 ± 0.7 | 45.7 |

The residual load bearing capacity of steel reinforced concrete beam is about 90% of initial value. The residual load bearing capacity of CFRP reinforced beam is about 66% of initial value, otherwise the beam reinforced with diameter 10 mm GFRP has the residual strength reduce to 31% of initial value.

The resin used for CFRP and GFRP rebars is the same. It can be assumed that the difference in relative residual strength of GFRP a CFRP is caused by the nature of glass fiber. Glass is generally considered as supercool liquid unlike to carbon, which is considered as thermally stable material.

## 4. Conclusion

Initial and residual load bearing capacity of concrete beams reinforced with four different reinforcements was determined. As a reinforcing material were used steel rebar of diameter 12 mm, carbon fiber reinforced polymer rebar of diameter 9 mm, and glass reinforced polymer rebar of diameter 10 and 14 mm. Concrete cover was 25 mm. Beams were exposed to temperature 1000°C for 2 hours. After cooling the beams:

- The steel reinforced concrete beam reached 90% of initial load bearing capacity.
- The carbon reinforced concrete beam reached 66% of initial load bearing capacity.
- The glass rebar of diameter 10 mm reinforced concrete beam reached 31% of initial load bearing capacity.
- The glass rebar of diameter 14 mm reinforced concrete beam reached 46% of initial load bearing capacity.

The slopes of load-displacement diagram are reduced in all cases (included the steel). This indicates that reduction in rebar to concrete bond occurred. It can be caused by irreversible temperature change in concrete.

The temperature measured on the reinforcement in the mid-span (zone exposed to 1000°C) reach critical temperature 325°C in 60 minutes or later. The temperature measured in anchorage zone ("cold end of the beam") reach 120°C (Tg of resin) in 150 minute or later. The residual load bearing capacity is 31–66%. This indicates that fire resistance of such beam can reach at least 60 minutes according the standard [8]. FRP reinforcement can be considered as safe fire-proof solution, although diameter and material of FRP should be taken into account.

The increasing Tg of resin to 150°C or more can rise up the fire resistance to theoretical limit 180 minutes.

Future research will be focused on the elevated temperatures effect on the residual behaviour of concrete beams reinforced with both GFRP rebars and stirrups.

**Acknowledgments**

This research was supported by the Technology Agency of the Czech Republic, grant No. TH04020431 (Extension of the application area of FRP reinforcement in concrete structures).

**References**

[1] Wang X, Zha X and Zhang X 2013 Bond behavior of FRP rebar and concrete at elevated temperature *Harbin Gongye Daxue Xuebao/Journal of Harbin Institute of Technology* **45** pp 8–15

[2] Hamad R, Megat J, Megat A and Haddad R 2017 Mechanical properties and bond characteristics of different fiber reinforced polymer rebars at elevated temperatures *Construction and Building Materials* **142** pp 521–535 DOI: 10.1016/j.conbuildmat.2017.03.113.

[3] Ellis D, Tabatabai H and Nabizadeh A 2018 Residual Tensile Strength and Bond Properties of GFRP Bars after Exposure to Elevated Temperatures *Materials* **11**(3) DOI: 10.3390/ma11030346.

[4] Özkal F, Polat M, Yağan M and Öztürk M 2018 Mechanical properties and bond strength degradation of GFRP and steel rebars at elevated temperatures *Construction and Building Materials* **184** pp 45–57 DOI: 10.1016/j.conbuildmat.2018.06.203.

[5] Daněk P, Rozsypalová I and Karel O 2019 Experimental study of concrete beams reinforced with GFRP rebars exposed to high temperatures *Key Engineering Materials* **808** pp 177–182 DOI: 10.4028/www.scientific.net/KEM.808.177.

[6] ISO 834 standard serie: Fire-resistance tests – Elements of building construction 1999 Technical Committee ISO/TC 92

[7] CSA - S806 (2012) - Design of Buildings with Fibre Reinforced Polymers

[8] ČSN EN 13501-2: Fire classification of construction products and building elements - Part 2: Classification using data from fire resistance tests, excluding ventilation services 2004 (Czech office for standards, metrology and testing)

# Comparison of structure and properties of gypsum mortars with different types of aggregates

**M Dolezelova**[1]**, J Krejsova**[1]**, L Scheinherrova**[1] **and A Vimmrova**[1]

[1] Department of Materials Engineering and Chemistry, Faculty of Civil Engineering, Czech Technical University in Prague, Thákurova 7, 166 29 Prague, Czech Republic

Email: vimmrova@fsv.cvut.cz

**Abstract**. Gypsum is one of the most environmentally friendly building materials and therefore its importance increases nowadays. Since the gypsum is mostly used in the form of the paste, the influence of aggregates on properties of gypsum mortars is not investigated very often. The article deals with the structure and properties of gypsum mortars with two different aggregates (silica sand and basalt). It was found, that even if the chemical composition and shape of particles of both aggregates differs significantly, the structure and properties of gypsum mortars with both types of aggregates are similar and the properties of the gypsum composite depends more on the aggregate granulometry and the quality of its surface. The aggregate with rougher surface of particles has more distinct interfacial transition zone, higher porosity and higher compressive strength.

## 1. Introduction

Even if gypsum is one of the oldest building binders, the knowledge of its structure, properties and behavior is very often based only on the historical experience or on the knowledge obtained from the study of other binders. While in the cement- and lime-based materials the presence of aggregates is inevitable, because they provide better dimensional stability at setting [1], gypsum does not shrink during setting; to the contrary it expands slightly [2]. Therefore the aggregates in gypsum are used less frequently and mostly from economic reasons. They could also improve the behavior of gypsum-based materials in fire [3, 4]. Even if the gypsum is considered as an excellent fire resistant material, gypsum paste shrinks considerably at high temperatures and the presence of aggregates can improve the volume stability of gypsum based material significantly [4].

It is known that the size, shape and texture of the aggregate particles influence the properties of the cement-based material. The properties of the aggregates affect the workability of the fresh mixture, the mechanical properties and the durability of the hardened cement mortars or concretes [1]. It was verified that the properties of aggregates influence the gypsum-based materials in similar way as cement-based materials [5].

Nevertheless the influence of normal types of aggregates on the gypsum materials is not studied very often. Mostly the lightweight aggregates are tested [6, 7, 8, 9]. Also large scale of waste products was used in gypsum-based materials, e.g. recycled expanded polystyrene [10], waste PUR [11] or ground rubber [12, 13].

Gypsum mortars with two different types of aggregates (silica sand and basalt) were studied. The mortar with silica sand was used as a reference material and basalt was chosen in order to improve the fire behavior of gypsum mortar. Basic physical and structural properties of both mortars were tested and compared in the paper.

## 2. Materials and methods

The commercial gypsum plaster (calcium sulfate hemihydrate, $CaSO_4 \cdot 1/2H_2O$) was used as a binder. Plaster was type A2, according ČSN EN 13 279-1:2009 [14] and it was calcined from flue gas desulfurization product ($CaSO_4 \cdot 2H_2O$) by producer Saint-Gobain Construction Products CZ, branch RIGIPS, Czech Republic.

Silica sand, which complies with CEN, ČSN EN 196-1 [15] and crushed basalt (producer KÁMEN Zbraslav, ltd.) were used as a fine aggregates. Set retarder Retardan-200 P (producer SIKA, Germany) was used in all mortars.

The chemical composition of raw materials, obtained by XRF analysis and their mineral composition, obtained by XRD analysis are given in the table 1 and 2. The granulometry of the aggregates can be seen in figure 1 and their bulk density and specific gravity is given in table 3. The bulk density and specific gravity of basalt are slightly higher than values of silica sand, but the difference was not significant. The granulometries of both aggregates obtained by the sieve analysis were similar, silica sand was slightly rougher. Nevertheless there was significant difference in the size and amount of smallest particles, obtained by laser analysis. Crushed basalt contained considerably larger amount of dust particles smaller than 20 μm (figure 2).

**Table 1.** Chemical composition of raw materials.

| Oxide | $Na_2O$ | $MgO$ | $Al_2O_3$ | $SiO_2$ | $SO_3$ [wt.%] | $K_2O$ | $CaO$ | $TiO_2$ | $Fe_2O_3$ |
|---|---|---|---|---|---|---|---|---|---|
| Gypsum | - | - | 0.4 | 0.6 | 53.9 | - | 44.5 | - | - |
| Standard sand | - | - | 1.1 | 97.7 | - | - | 0.2 | - | 0.4 |
| Basalt | 3.3 | 7.3 | 17.2 | 42.2 | - | 1.2 | 12.7 | 3.2 | 11.6 |

**Table 2.** Mineral composition of raw materials.

| | Mineral | Gypsum [wt. %] | Silica sand [wt. %] | Basalt [wt. %] |
|---|---|---|---|---|
| Bassanite | $CaSO_4 \cdot 1/2H_2O$ | 89 | - | - |
| Anhydrite | $CaSO_4$ | 6 | - | - |
| Calcite | $CaCO_3$ | 4 | - | - |
| Muscovite | $KAl_2(AlSi_3O_{10})(OH)_2$ | 1 | - | 11 |
| Quartz | $SiO_2$ | - | 100 | - |
| Diopside | $CaMgSi_2O_6$ | - | - | 44 |
| Nepheline | $(Na,K)AlSiO_4$ | - | - | 8 |
| Plagioclase | $NaAlSi_3O_8 - CaAl_2Si_2O_8$ | - | - | 10 |
| Anorthite | $CaAl_2Si_2O_8$ | - | - | 2 |
| Analcim | $NaAlSi_2O_6 \cdot H_2O$ | - | - | 5 |
| Forsterite | $Mg_2SiO_4$ | - | - | 6 |
| Aegirine | $NaFe^{3+}[Si_2O_6]$ | - | - | 2 |
| Tobermorite | $Ca_5Si_6O_{16}(OH)_2 \cdot 4H_2O$ | - | - | 3 |
| Biotite | $K(Mg,Fe)_3(AlSi_3O_{10})(F,OH)_2$ | - | - | 3 |
| Magnetite | $Fe^{2+}Fe^{3+}{}_2O_4$ | - | - | 4 |

**Table 3.** Bulk density and specific gravity of aggregates.

| | | Silica sand | Basalt |
|---|---|---|---|
| Bulk density | [kg m$^{-3}$] | 2575 | 2631 |
| Spec. gravity | [kg m$^{-3}$] | 2831 | 3015 |

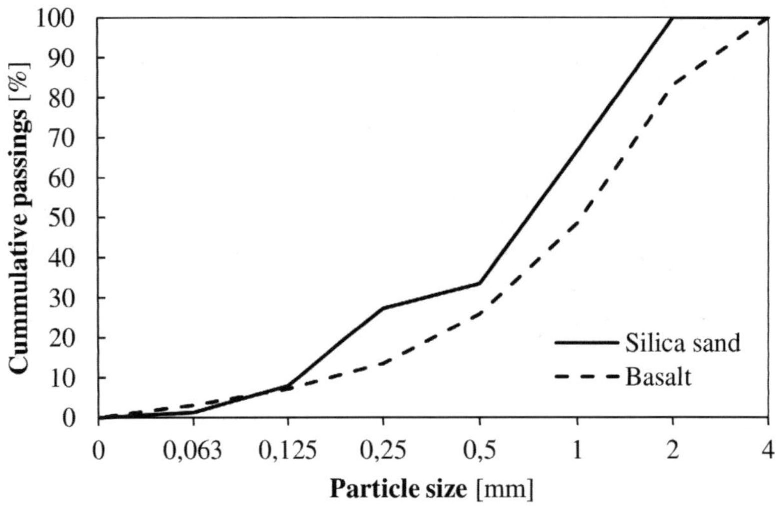


**Figure 1.** Granulometry of used aggregates.

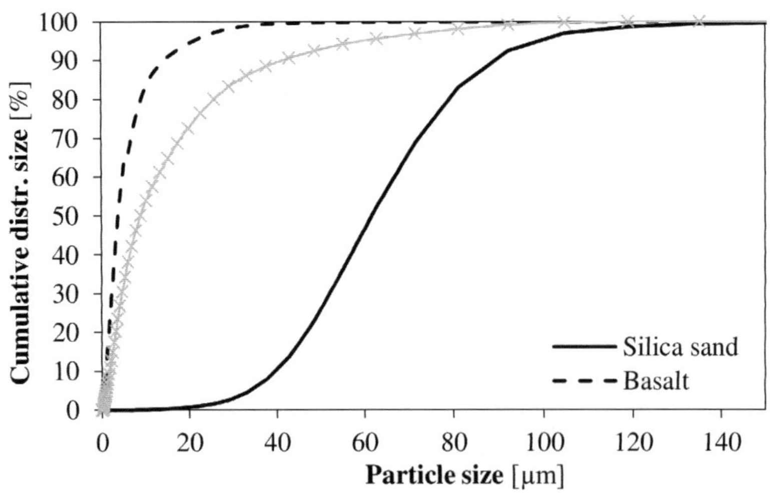


**Figure 2.** Results of laser particle size analysis.

*2.1. Composition and preparation of mortars*

The composition of the mortars was designed for the same volume of gypsum and aggregate in both mortars in order to compare the structural changes in the gypsum matrix. Composition of mortar with silica sand (GS) was designed according ČSN EN 196-1 [15]. The mass ratio between the gypsum powder and sand was 1 : 3. Amounts of 450 g of gypsum and 1350 g of sand were used for one batch of mortar. The mass of basalt aggregate in mortar GB was calculated from its density $\rho_B$ and the volume of silica sand $V_S$ in mortar GS. By that way both mortars contained the same volume of aggregate.

The amount of water was determined for a flow tests diameter value of 165±5 mm. Amount of water in GB mortar was higher because basalt contained higher amount of very fine particles and therefore more water was needed to achieve same workability. The amount of retarding agent was 0.02% from dry gypsum weight in both mortars. Composition of tested materials is given in table 4.

**Table 4**. Composition of mortars.

| Material | Aggregate | Gypsum | Aggregate | Retardan | Water |
|---|---|---|---|---|---|
| | | | [g/batch] | | [ml/batch] |
| GS | silica sand | 450 | 1350.0 | 0.09 | 270 |
| GB | basalt | 450 | 1379.4 | 0.09 | 330 |

The materials were prepared according to ČSN EN 13454-2 [16]. Dry gypsum powder was mixed with the retarding agent and dry mixture was poured into the required amount of water in the mixing bowl and the bowl was put into the automatic mixer. Total time of mixing was 4 minutes. Mixing consisted from 30 s at low speed (140 rpm), 30 s of adding aggregates at low speed, 30 s mixing at high speed (285 rpm), pause 90 s during which the mixture was manually wiped from the bowl wall and finally mixing for 60 s at high speed. Then the mortar was poured into a mold and was firstly manually and then mechanically compacted. Three prismatic samples (40 × 40 × 160 mm) were prepared from each mortar. The hardened samples were put out of the molds after 60–90 min and stored at 20 ± 5°C and relative humidity of 50 ± 5% for 28 days.

### *2.2. Methods*

The **particle size distribution** of used aggregates was determined according to ČSN EN 933-2 [17]. Standard sieves with apertures of 0.063 mm, 0.09 mm, 0.125 mm, 0.25 mm, 0.5 mm, 1.0 mm, 2.0 mm and 4.0 mm were used.

The **bulk density** of aggregates was determined by the immersion of aggregate in the volumetric cylinder and counted from dry mass of aggregate and the increase of water level in volumetric cylinder. This method is not standardized.

The bulk density of mortars was calculated from the mass and dimensions as a ratio of mass and volume. The dimensions of samples were measured by the digital calliper and their mass was determined by weighing. Volume of samples was calculated from dimensions.

The **specific gravity** of aggregates and tested mortars was determined by the helium pycnometry using Pycnomatic ATC (Thermo Fisher Scientific). This method uses a gas displacement technique and it is not standardized.

The **microstructural morphology** of the gypsum mortars and aggregate grains was studied by scanning electron microscopy (SEM) using a Phenom XL electron microscope. The samples were dried and they were not coated or polished.

The **grain and fracture surface roughness** was determined by the 3D Roughness Reconstruction software, based on "shape from shading" technology from SEM images, obtained by the device Phenom XL. Roughness was determined as the three-dimensional arithmetical mean roughness value SRa [μm], which is an analogy to the arithmetical mean deviation of the profile Ra (two dimensional roughness) according to ISO 468 [18].

The porosity of mortars p [%] was calculated from the bulk density $\rho_v$ and the specific mass $\rho$ according equation 1.

$$p = (1 - \rho_v/\rho) \times 100 \quad (1)$$

Mercury intrusion porosimetry (MIP) was used for describing of **pore system** of studied composites. The measuring apparatus was composed of devices Pascal 140 and Pascal 440. The software evaluated the cumulative curve of pores and the distribution curve of pores.

The **flexural and compressive strength** were determined according ČSN EN 13454-2 [16] on standard test samples. The experiment was performed as a common three-point bending test. The measurements were carried out 28 days after mixing. The compressive strength was measured on the halves of the specimens left over from the bending test.

The **modulus of elasticity** was measured according ČSN EN 12504-4 [19] by the Pundit Lab Ultrasonic device (Proceq) using the ultrasonic method. The modulus of elasticity was calculated from bulk density $\rho_v$ and pulse velocity v according equation 2.

$$E = \rho_v \times v^2 \quad (2)$$

## 3. Results and discussion

Basic properties of both mortars GS and GB are given in table 5 together with values of surface roughness of both aggregates and the fracture surface roughness of mortars. The values of all properties are discussed in the following chapters.

**Table 5.** Properties of tested mortars.

| | | GS | GB |
|---|---|---|---|
| Specific gravity | [kg m$^{-3}$] | 2754 | 2938 |
| Bulk density | [kg m$^{-3}$] | 1916 | 1903 |
| Total porosity | [%] | 30.4 | 35.2 |
| Compressive strength | [MPa] | 12.7 | 13.8 |
| Flexural strength | [MPa] | 5.9 | 5.0 |
| Modulus of elasticity | [MPa] | 3434 | 2936 |
| Grain surface roughness | [µm] | 591 | 633 |
| Fracture surface roughness | [µm] | 0.30 | 0.37 |
| Thermal conductivity | [W m$^{-1}$ K$^{1}$] | 1.8 | 0.7 |

### *3.1. XRD analysis of mortars*

XRD analysis showed that used aggregates did not react with the gypsum matrix. The mineral composition of both mortars contained the same minerals, which were found in the raw materials, only bassanite (calcium sulphated hemihydrate) in calcined gypsum powder was hydrated to the gypsum (calcium sulphate dihydrate). It is evident that both aggregates were chemically inert in gypsum environment. The mineral composition of both mortars is given in table 6.

**Table 6.** Mineral composition of mortars.

| | | GS | GB |
|---|---|---|---|
| Mineral | | [wt.%] | |
| Gypsum | $CaSO_4 \cdot 2H_2O$ | 38 | 46 |
| Bassanite | $CaSO_4 \cdot 1/2H_2O$ | 1 | 2 |
| Anhydrite | $CaSO_4$ | - | 0 |
| Muscovite | $KAl_2(AlSi_3O_{10})(OH)_2$ | - | 5 |
| Quartz | $SiO_2$ | 61 | - |
| Diopside | $CaMgSi_2O_6$ | - | 22 |
| Nepheline | $(Na,K)AlSiO_4$ | - | 4 |
| Plagioclase | $NaAlSi_3O_8 - CaAl_2Si_2O_8$ | - | 4 |
| Anorthite | $CaAl_2Si_2O_8$ | - | 2 |

| | | | |
|---|---|---|---|
| Analcim | $NaAlSi_2O_6 \cdot H_2O$ | - | 2 |
| Forsterite | $Mg_2SiO_4$ | - | 3 |
| Aegirine | $NaFe^{3+}[ Si_2O_6]$ | - | 2 |
| Tobermorite | $Ca_5Si_6O_{16}(OH)_2 \cdot 4H_2O$ | - | 0 |
| Biotite | $K(Mg,Fe)_3(AlSi_3O_{10})(F,OH)_2$ | - | 2 |
| Magnetite | $Fe^{2+}Fe^{3+}{}_2O_4$ | - | 2 |

### *3.2. SEM observations and surface roughness evaluation*

The shape and quality of the surface of the aggregate grains were studied by the SEM. From each aggregate several grains were chosen and the surface roughness SRa was determined 10 times for each aggregate at 20 μm wavelength and the mean value was calculated. SEM images of typical grain surface and height map generated by 3D Roughness Reconstruction software can be seen in figure 3.

Both aggregates were crushed, therefore their surface was not smooth, nevertheless the basalt surface was rougher and there can be seen (figure 4), that the ITZ between gypsum paste and aggregates was more pronounced in the GB mortar. The size and shape of gypsum crystals in the vicinity of aggregate particles were similar in both mortars

a)

b)

**Figure 3.** SEM image at 1000× magnification (a) and height map (b) of typical grains.

GS GB

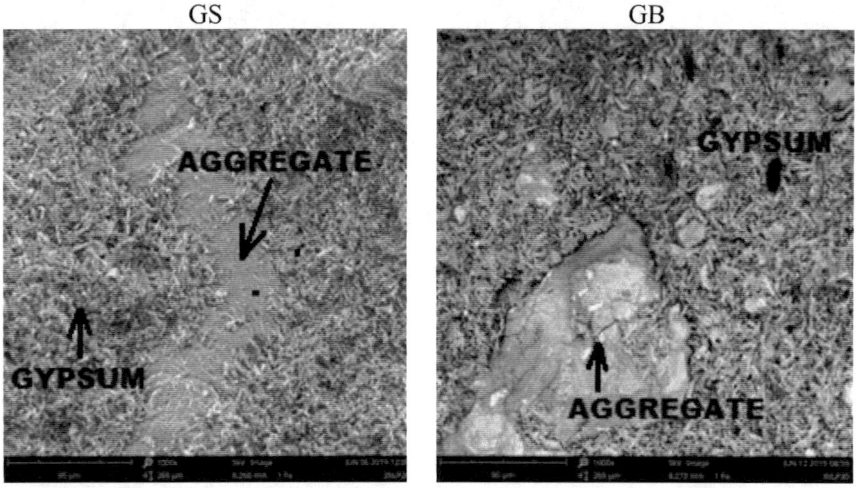


a)

b)

**Figure 4**. SEM image of mortars at 1000× (a) and 3000× (b) magnification.

### *3.3. Porosity and pore size distribution of mortars*

The total porosity of mortar with silica sand GS was 30.4%, porosity of mortar with basalt GB was 35.2%. Even if the specific gravity of mortar with basalt aggregate (GB) was higher than specific gravity of mortar with silica sand (GS), its bulk density was lower, because it contained more water.

The increase of porosity of GB mortar was therefore caused mainly by the higher amount of water and partly by larger ITZ in this mortar (as can be seen in SEM images). The size of more porous ITZ increases with the roughness of the aggregate particles [5].

The pore size distribution curves are given in figure 5. The distribution curves of both mortars are similar. They have nearly unimodal pore size distribution with highest amount of pores between 1–10 μm, which is typical for gypsum pastes [20]. The higher amount of pores larger than 10 μm in mortar GB was caused by the higher amount of water. Mortar GB contained also more very small pores (< 0.01 μm), which correspond with the fact that smaller pores appears in the gypsum mortars with rougher surface of particles [5].

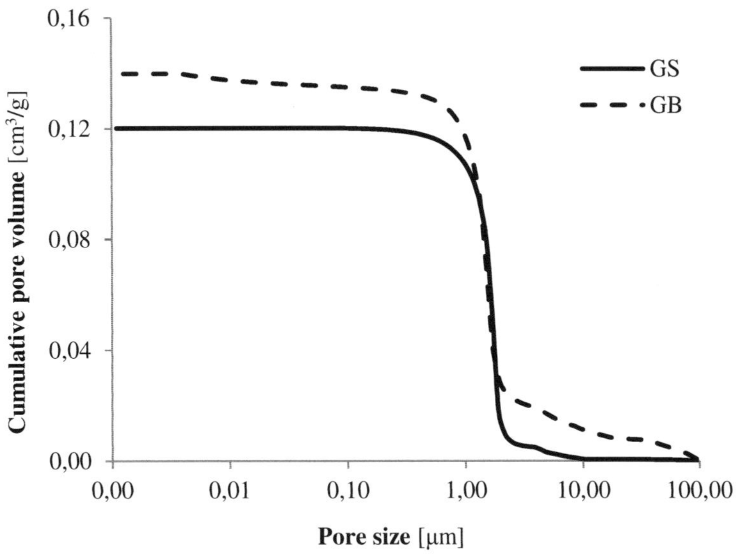


**Figure 5.** Pore size distribution curve of mortars.

### *3.4. Mechanical and thermal properties of mortars*

The compressive strength of mortar GS was 12.8 MPa and mortar GB had the compressive strength 13.2 MPa. The higher compressive strength of mortar GB with basalt was caused by the higher surface roughness of basalt particles compared to the silica sand. This corresponds with the conclusions in the [5] and also with the behaviour of cement based materials [1]. The higher porosity of the mortar GB had no impact on the compressive strength, because the effect of surface roughness prevailed. The flexural strength was more influenced by the porosity and therefore the flexural strength of GB mortar was about 15% lower than flexural strength of mortar GS.

The modulus of elasticity decreased with increasing porosity (from 3434 MPa for GS to 2936 for GB), which is common for porous materials [21, 22].

The thermal conductivity of mortar with basalt GB decreased significantly (at about 50%), compared to the thermal conductivity of mortar with silica sand GS, which conforms to the higher porosity of the mortar GB.

## 4. Conclusions

Mortars with two different types of nonporous aggregates were studied. The mortar with silica sand was used as a reference material. Basalt was used in order to improve the fire behaviour of gypsum mortar. The comparison of mortars with both types of the aggregates was performed at normal temperature before subjecting them to the elevated temperatures.

It was found, that both aggregates did not chemically react with the gypsum matrix at normal temperature and the differences in the structure and properties of both mortars were caused by the aggregate particle size distribution and the quality of particles surface.

We found that:

- the shape and size of crystals of gypsum matrix were similar in both mortars
- the ITZ in the mortar with basalt was more pronounced, which was caused by the higher roughness of basalt surface
- the pore size distributions of mortars was unimodal with most of the pores having diameter 0.1–1 μm
- higher porosity of mortar with basalt aggregate was caused by higher amount of mixing water, caused by higher amount of very fine particles in basalt aggregate
- the thermal conductivity of mortar with basalt aggregate was lower at about 50%

The research will continue by the testing of both materials at the elevated temperatures (up to 1000°C). The behaviour of mortar with silica sand will be compared with the behaviour of the mortar with basalt aggregate and also with the behaviour of gypsum paste. Silica sand is generally considered as less suitable for use at high temperatures, because of its volume instability [23] and therefore its substitution by other material could be beneficial for utilization of gypsum based materials as a fire protecting materials.

**Acknowledgments**

This research was supported by the Czech Science Foundation, Project No. 19-08605S.

**References**

[1] Mindess S, Young J F and Darwin D 2003 *Concrete* (Upper Saddle River (NJ): Prentice Hall)

[2] Karni J and Karni E 1995 Gypsum in construction: origin and properties *Mater. Struct.* **28** 92

[3] Féjean J, Lanos C, Mélinge Y and Baux C 2003 Behaviour of Fire-Proofing Materials Containing Gypsum, Modifications Induced by Incorporation of Inert Filler *Chem. Eng. Res. Des.* **81** 1230

[4] Doleželová M, Scheinherrová L, Krejsová J and Vimmrová A 2018 Effect of high temperatures on gypsum-based composites *Constr. Build. Mater.* **168** 82

[5] Krejsová J, Doleželová M, Pernicová R, Svora P and Vimmrová A 2018 The influence of different aggregates on the behavior and properties of gypsum mortars. *Cem. Concr. Compos.* **92** 188

[6] Gencel O, del Coz Diaz J, Sutcu M, Koksal F, Álvarez Rabanal F and Martínez-Barrera G 2016 A novel lightweight gypsum composite with diatomite and polypropylene fibers *Constr. Build. Mater.* **113** 732

[7] Hernández-Olivares F, Bollati M, del Rio M and Parga-Landa B 1999 Development of cork–gypsum composites for building *Constr. Build. Mater.* **13** pp 179–186

[8] García Santos A 2009 Escayola reforzada con fibras de polipropileno y aligerada con perlas de poliestireno expandido *Materiales de Construcción* **59**

[9] Demir I and Serhat Baspinar M 2008 Effect of silica fume and expanded perlite addition on the technical properties of the fly ash–lime–gypsum mixture *Constr. Build. Mater.* **22** 1299

[10] González Madariaga F 2008 EPS (expanded poliestyrene) recycled bends mixed with plaster or stucco, some applications in building industry *Informes de la Construcción* **60**

[11] Gutiérrez-González S, Gadea J, Rodríguez A, Blanco-Varela M and Calderón V 2012 Compatibility between gypsum and polyamide powder waste to produce lightweight plaster with enhanced thermal properties *Constr. Build. Mater.* **34** 179

[12] Herrero S, Mayor P and Hernández-Olivares F 2013 Influence of proportion and particle size gradation of rubber from end-of-life tires on mechanical, thermal and acoustic properties of plaster–rubber mortars *Mater. Des.* **47** 633
[13] Jiménez Rivero A, de Guzmán Báez A and García Navarro J 2014 New composite gypsum plaster – ground waste rubber coming from pipe foam insulation *Constr. Build. Mater.* **55** 146
[14] ČSN EN 13 279-1 2014 Gypsum binders and gypsum plasters. Definitions and requirements (Prague: Czech Standardization Institute)
[15] ČSN EN 196-1 Methods of testing cement – Part 1: Determination of strength 2016 (Prague: Czech Standardization Institute)
[16] ČSN EN 13454-2 Binders, composite binders and factory made mixtures for floor screeds based on calcium sulfate – Part 2: Test methods 2019 (Prague: Czech Standardization Institute)
[17] ČSN EN 933-2 Tests for geometrical properties of aggregates. Determination of particle size distribution. Test sieves, nominal size of apertures 1997 (Prague: Czech Standardization Institute)
[18] ISO 468. Surface roughness. Parameters, their values and general rules for specifying requirements 1982 (International Organization for Standardization)
[19] ČSN EN 12504-4 Testing concrete. Determination of ultrasonic pulse velocity 2005 (Prague: Czech Standardization Institute)
[20] Tesárek P, Drchalová J, Kolísko J, Rovnaníková P and Černý R 2007 Flue gas desulfurization gypsum: Study of basic mechanical, hydric and thermal properties *Constr. Build. Mater.* **21** 1500
[21] Luo J and Stevens R 1999 Porosity-dependence of elastic moduli and hardness of 3Y-TZP ceramics *Ceram. Int.* **25** 281
[22] Kováčik J 1999 Correlation between Young's Modulus and Porosity in Porous Materials *J. Mater. Sci. Lett.* **18** 1007
[23] Sorrell C, Anderson H and Ackermann R, 1974. Thermal expansion and the high–low transformation in quartz. II. Dilatometric studies. *J. Appl. Crystallogr.* **7** 468

# HPC cement materials prepared by mixing under reduced pressure

**J Vojtisek[1], R Novotny[1], E Bartonickova[1], P Hruby[1], M Marko[1] and M Janca[1]**

[1] Brno University of Technology, Faculty of Chemistry, Purkynova 118, 612 00 Brno, Czech Republic

Email: xcvojtisek@fch.vut.cz

**Abstract.** This article deals with HPC (High Performance composite) materials whose main advantages are increased mechanical properties, durability and stability. The limiting parameter for such materials is the presence of air bubbles that cannot be removed just by the compacting. Mixing of cement mixtures under reduced pressure can be used to reduce porosity and eliminate the presence of air bubbles. Moreover, it is associated by the improvement of the mechanical properties. The influence of preparation of HPC by mixing under the atmospheric pressure and by mixing under reduced pressure to parameters like mechanical properties, porosity, microstructure was studied in this paper. The porosity of the samples was analysed in a cross-section using an optical method. The effect of the reduced pressure on the resulting strength of the samples was observed.

## 1. Introduction

Concrete is the most common building material elsewhere on the world. The typical concrete usually consists of Ordinary Portland Cement (OPC) mixed with water and coarse aggregates which is accompanied with a low homogeneity. The nowadays society focuses not only on the utility properties of the materials but on the environmental and economic aspects as well. The clinker production process is related with rapid $CO_2$ emission (about 7% of total world´s production), thus it is not surprising that scientists all around the world are trying to find some alternatives, whose can help to eliminate production of greenhouse gases. The most common is probably utilization of secondary raw materials. Considering their utilization (blast furnace slag, fly ash, microsilica) as a partial replacement of clinker in cements improvement of mechanical properties, durability same as the reduction of the size of the particles can be achieved [1].

For the same amount or reduction of Portland cement and increase of mechanical properties of the material, it is possible to reduce the size of the particles or add reactive components (such as blast furnace slag, microsilica, fly ash). Another important factor, which influences mechanical parameters of concrete, is the water to binder ratio (w/b). The reduction of water is achieved by plasticizing additives. These types of concrete are typically classified as HPCs (high-performance cement composites). Concrete prepared from a large amount of Portland cement is characterized by a high hydration temperature. This cause in micro-cracks. Reduction of microcracks formed in concrete due to this effect can be achieved by water curing [1, 2].

RPC (Reactive powder concrete) is a new field of development of cement materials and it belongs to the UHPC (Ultra – high performance composite) category. UHPC materials were developed due to various aspects. The main ones were high price of OPC and environmental burden associated with the cement production beside the superior mechanical properties and durability when compared with traditional concrete based on OPC. The RPCs allow us to produce material with improved mechanical properties, with lower price and environmental burden than the traditional concrete (with OPC). Instead of traditional aggregates, very fine sand is used in this case. Fine sand with a maximum particle size of 0.4 mm is used. The low w/b is compensated by the addition of a superplasticizer. RPC materials use very small highly homogenous particles without aggregates to form dense and homogenous matrix. Due to these properties, micro-cracks are limited and shrinkage is very small. It is possible to achieve compressive strengths around 200 MPa after 28 days [2, 3, 4, 5, 6].

The pozzolanic reaction of this material may be similar to traditional compositions. The hydration processes depend on the type of curing (dry, water) same as on the type of reactive component of concrete. An undesirable property of this material is increased brittleness. Fibres (steel, carbon, PVA) can be doped into the matrix to increase the toughness and material durability [1, 7, 8].

The microstructure is an important factor in concrete preparation. To improve the microstructure, it is possible to realize aging of concrete at elevated temperatures or to improve the mixing process. Mixing under reduced pressure causes a high reduction of the porosity. Porosity can cause local stress concentration which can be accompanied by the material rupture. Vacuum mixing can limit these problems. The porosity of concrete with standard composition is around 5–10%. In samples under reduced pressure, it is possible to achieve 1/10 porosity [9, 10, 11, 12].

This article deals with the preparation of RPC materials with the aim to compare two mixing methods – standard mixing and mixing under reduced pressure. The influence of the mixing procedure on the microstructure, porosity and mechanical parameters was studied and evaluated. Reduced porosity can be expected to increase mechanical resistance.

## 2. Materials and experimental part

White Portland cement CEM I 52.5 R (Aalborg), silica fume RW-Füller Q1 (RW silicium, Germany), limestone SBL CL 90-Q (Kotouč Štramberk, spol. s r.o., Czech Republic), fine sand (Filtrační písky, spol. s r.o., Czech Republic), superplasticizer Master Glenium ACE 446 (BASF Stavební hmoty Česká republika s.r.o., Czech Republic) and deionised water were used in this study. The chemical composition of cement is given in table 1, the particle size distribution of sand, limestone, silica fume and cement shows table 2.

Chemical composition was measured by manufacturer by Heidelberg Cement group and it is shown in table 1.

**Table 1.** Chemical composition of the applied white cement determined by XRF.

| | $C_3S$ | $C_2S$ | $C_3A$ | $C_4AF$ | MgO | $SO_3$ | Cl | $Na_2O$ |
|---|---|---|---|---|---|---|---|---|
| CEM I 52.5 R Aalborg white | 73 | 16 | 5 | 1 | 0,6 | 1.8-2.3 | ≤0.04 | ≤0.3 |

Particle size distribution of CEM I 52.5 R Aalborg cement and silica fume measured by laser diffraction analyzer (HELOS KR Sympatec) are shown in table 2.

**Table 2.** Distribution of particle sizes.

| | D50 [μm] |
|---|---|
| CEM I 52.5 R Aalborg white | 8.7 |
| Quartz flour ST 09 | 6 |
| Fine sand | 140 |
| Silica fume | 1.9 |

The dry components were homogenized in the mixer (KitchenAid) for 1 minute. After that mix 75% of water and superplasticizer were added. The mixture was stirred for 5 minutes. After that 25% of the remaining water was added and stirred for 3 minutes. The mixture was transferred to molds with a test specimens´ dimension of 40 × 40 × 160 mm. These forms were compacted by vibrating table for 45 seconds.

The mixture was placed into the vacuum mixer and stirred intensively for 1 minute, before compaction in case of samples under the reduced pressure. Subsequently, the mixtures were placed in molds and compacted, as in the previous samples without the use of the vacuum. After 24 hours, the experimental specimens were demoulded and placed in temperature-controlled (laboratory temperature 25°C) water curing for optimal concrete aging.

Processability of the mixture was tested by Hägermann-flow table for the flow test. Firstly, filled cone was removed and spread was measured. Second test was measured after 15 strokes.

Mechanical properties were tested by DESTTEST, BETONSYSTEM (FCH BUT). Flexural strength (3-point arrangement) and compressive strength tests were performed after 1, 7 and 28 days. Furthermore, the density of concrete samples was measured. After 28 days, samples were sawed and the level of porosity was determined. A white cement paste was prepared to fill the pores. The prepared samples were photographed. The porosity was evaluated using an ImageJ program. The efficiency of mixing under the reduced pressure was evaluated based on the results obtained (compressive and flexural strengths; porosity).

## 3. Results and discussion

### *3.1. Preparation of RPC material*

The solid phase composition was the same for all mixture experiments. The highest water to binder ratio used was 0.22. Other samples were prepared with a 0.02 reduction step up to the workability limit. The mixture with a w/b = 0.22 was very fluid and easy to process. During forming, easy degassing was possible due to the fluidity of the system. The next mixture was w/b = 0.20, where the viscosity increased but the workability of the mixture was optimal. The last step was the mixture with w/b = 0.18, when the reduction of the amount of water in the mixture was ended, because the processability and the possibility to homogenize the mixture were difficult. Spread data are shown in table 3. The compositions of the individual mixtures are given in table 4.

**Table 3.** Summary of Hägermann-flow table test.

| | Before strokes [mm] | After 15 strokes [mm] |
|---|---|---|
| w/b 0.18 | 105 | 130 |
| w/b 0.2 | 105 | 145 |
| w/b 0.22 | 110 | 180 |

**Table 4.** Composition [g] of measured samples.

| | w/b 0.18 | w/b 0.20 | w/b 0.22 | w/b 0.18 vac | w/b 0.20 vac | w/b 0.22 vac |
|---|---|---|---|---|---|---|
| CEM I 52.5 R [g] | 225 | 225 | 225 | 225 | 225 | 225 |
| Silica fume[g] | 75 | 75 | 75 | 75 | 75 | 75 |
| Quartz flour ST_09 [g] | 75 | 75 | 75 | 75 | 75 | 75 |
| Fine sand [g] | 360 | 360 | 360 | 360 | 360 | 360 |
| Limestone SBL CL 90-Q [g] | 15 | 15 | 15 | 15 | 15 | 15 |
| Superplasticizer ACE 446 [g] | 7.5 | 7.5 | 7.5 | 7.5 | 7.5 | 7.5 |
| Deionised water [g] | 54 | 60 | 66 | 54 | 60 | 66 |

As shown in table 4, the same mixtures were stirred under reduced pressure. The mixture was transferred to a mixer for stirring under reduced pressure and stirred intensive for 1 minute. The pressure was measured with a barometer. The pressure in the mixer was adjusted to approximately 1/10 atmospheric pressure using an oil pump. After stirring, the mixture was homogeneously transferred to molds and compacted.

*3.2. Testing of mechanical parameters*

Samples were demoulded and put into a water curing after 24 hours to achieve a better final property of the material.

The flexural strength of the sample with w/b = 0.18 was about 8 MPa after 1 day of hydration. When compared to a sample under reduced pressure, influence of mixing under the reduced pressure on the increase in flexural strength was obvious. Same trend was observed for samples measured after 7 and 28 days. As expected, the maximum flexural strength was obtained by samples prepared using the mixing under the reduced pressure after 28 days of hydration. The flexural strengths measured were up to 32 MPa as shown in figure 1. The compressive strength of the samples was measured after 1 day without using vacuum mixing and reached 87 MPa. The maximum value (190 MPa after 28 days) was measured for the samples using reduced pressure mixing [9, 12, 13].

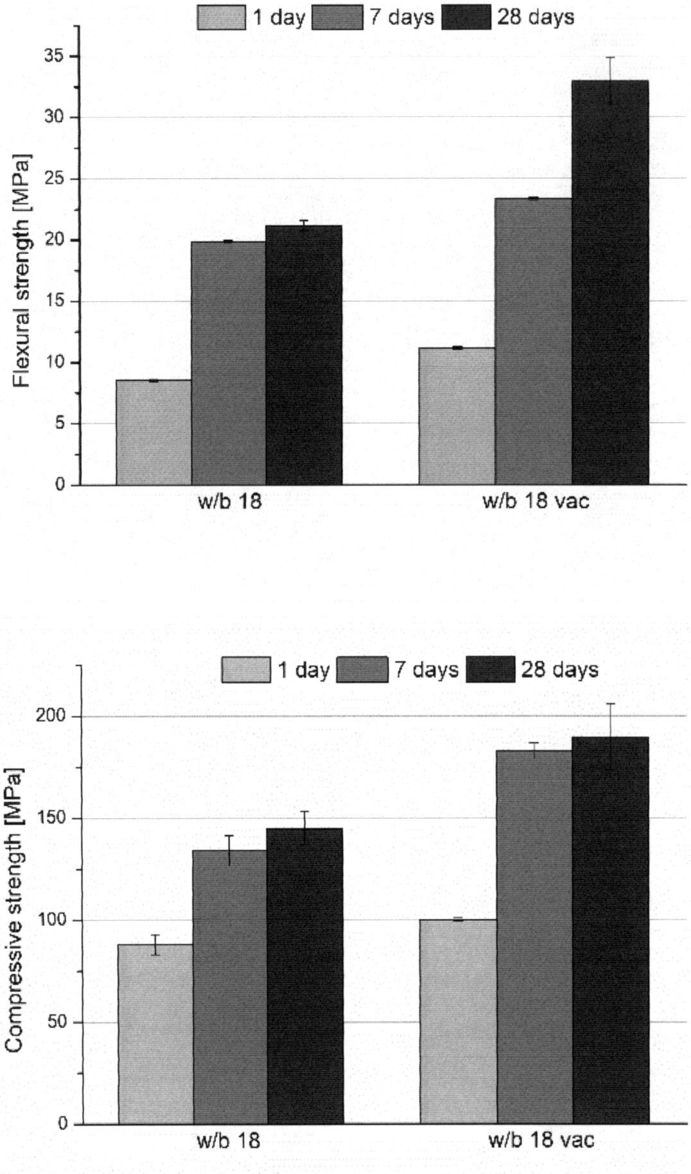


**Figure 1.** Flexural and compressive strength of the specimens w/b = 0.18.

The values of flexural and compressive strength of the specimens with w/b 0.18 are shown in figure 2. The obtained values were lower than the ones of mixtures with w/b = 0.2. It can be related to the higher w/b coefficient since the sufficiency of water is necessary to obtain the desired strength. The flexural strength of samples with w/b = 0.20 was 7 MPa after 1 day, which is not as much as in case of w/b = 0.18. The flexural and compressive strength after 28 days were 27 MPa, respectively 190 MPa for the samples prepared using the vacuum technique. No significant difference was observed between the samples with 18 and 20% water. The only difference was a better workability for the mixture with the 20% water content.

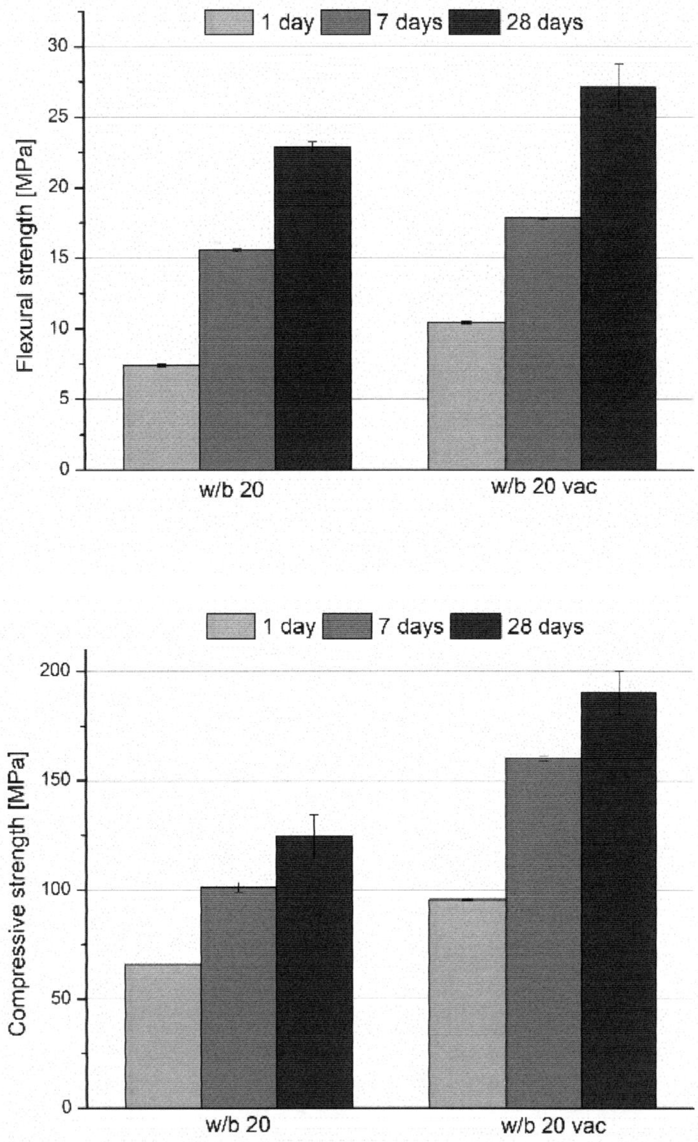


**Figure 2.** Flexural and compressive strength of the specimens w/b = 0.20.

Figure 3 summarizes the flexural and compressive strengths of specimens with w/b = 0.22. The reduction of compressive strength was observed in case of the specimen with w/b = 0.22 in comparison with specimens with w/b = 0.20 and 0.18. The compressive strength of the sample with w/b = 0.22 prepared under the reduced pressure is 160 MPa, which is 30 MPa less than in the case of the specimens with w/b = 0.20 and 0.18. This can be attributed to the relatively high water content in the mixture connected with excessive fluidity, which caused additional air capturing during the further processing (removing from the mixer, molding process) and thus the increase of porosity. Therefore, the optimal water coefficient for the tested mixtures was determined to be 0.2.

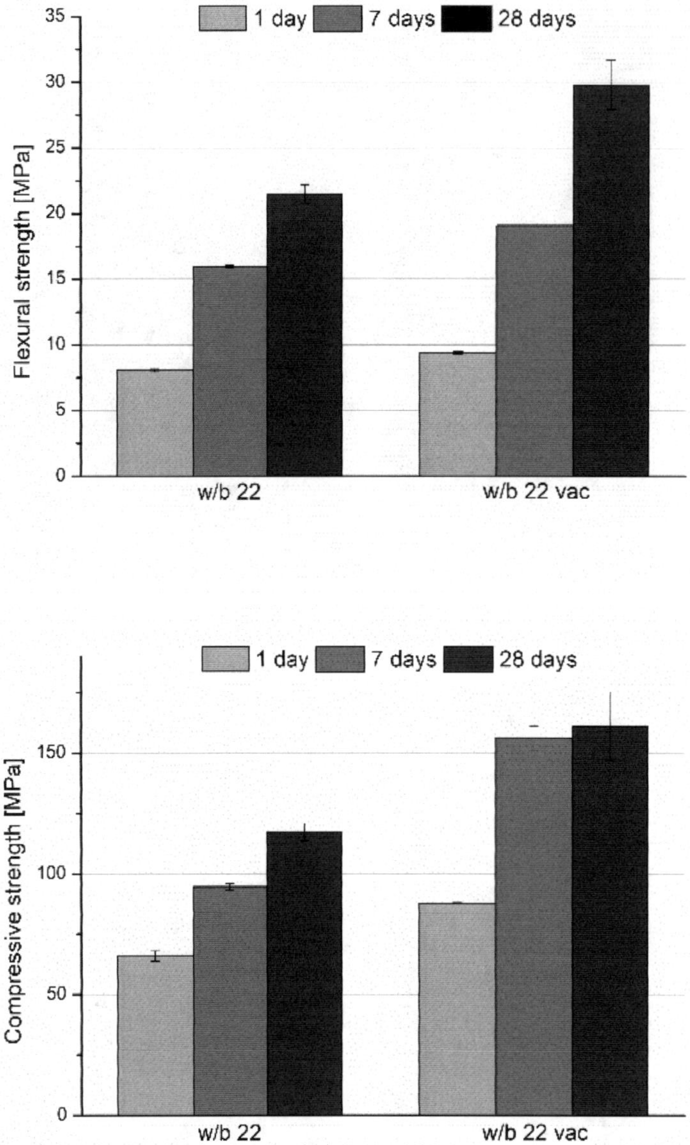


**Figure 3.** Flexural and compressive strength of the specimens w/b = 0.22.

### *3.3. Measure of porosity*

The porosity was significantly reduced by the mixing under the reduced pressure as can be seen in the figure 4. Figure 4 (a) is the sample prepared by mixing at atmospheric pressure where the pores were filled with white cement paste. Samples were photographed and analysed using ImageJ programme. The calculated porosity for the unvacuated samples with a water coefficient w/b = 0.18 was 7.3%, after stirring under reduced pressure only 0.67%. Such a significant decrease in porosity positively influences the mechanical resistance, which correlates with the data in figure 7 [9].

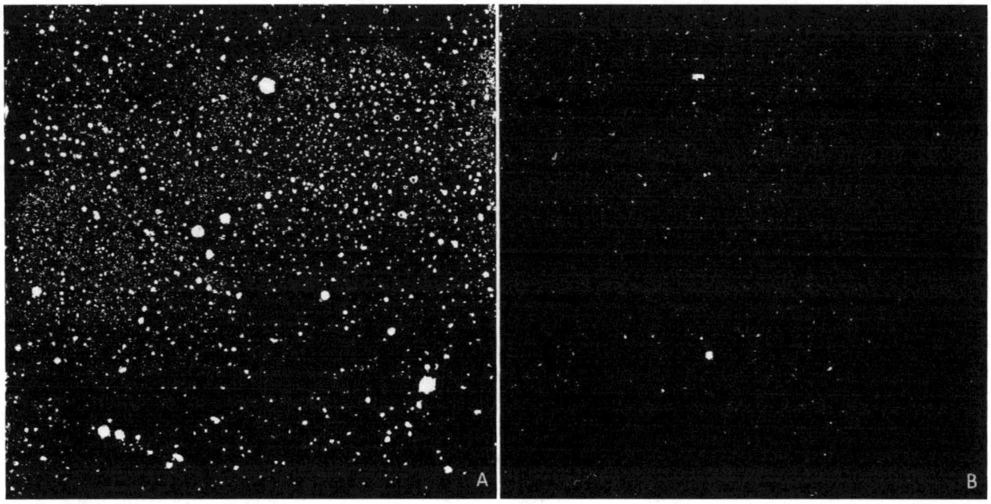


**Figure 4.** Analysis of porosity of the specimens w/b = 0.18, a) atmospheric pressure, b) reduced pressure.

The data in table 5 show similar level of porosity for specimen w/b = 0.20 and w/b = 0.18. Significant decrease of porosity of the samples was observed when the vacuum was applied during the mixing process, as can be seen in figure 5. An apparent positive influence of vacuum mixing on the physico-mechanical properties, in particular the strength and porosity, was also reported by [11].

**Figure 5.** Analysis of porosity of the specimens w/b = 0.20, a) atmospheric pressure, b) reduced pressure.

In the figure 6, a higher porosity is visible, when compared with the samples prepared under reduced pressure. In table 5, the porosity values are shown and it can be seen that the mixture was probably too fluid and thus new porosity was formed in the samples during moulding. This correlates with the measured mechanical strength of the specimen w/b = 0.22, which did not achieve as high resistance as the samples with lower water coefficient. Similar results were obtained by Zdeb T., who described a significant porosity reduction after mixing with reduced pressure [14].

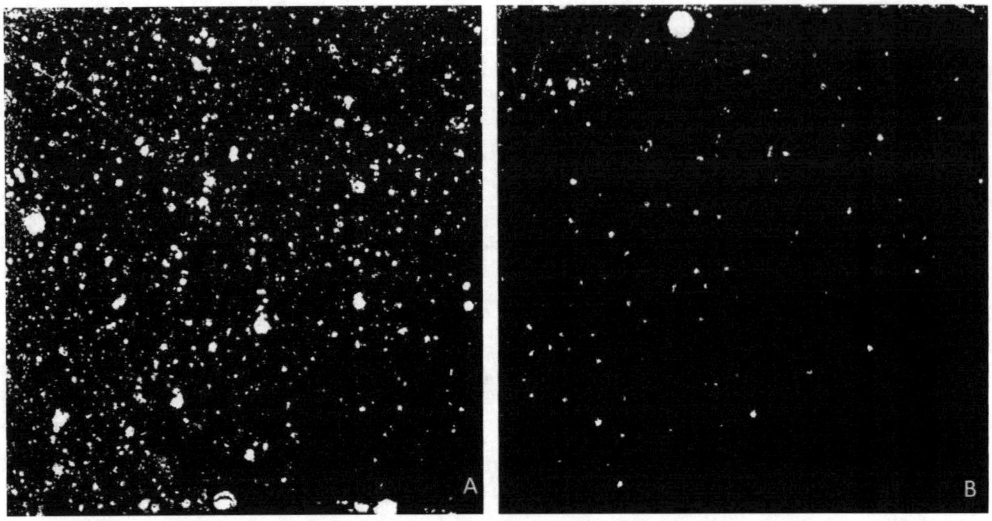


**Figure 6.** Analysis of porosity of the specimens w/b = 0.22, a) atmospheric pressure, b) reduced pressure.

**Table 5.** Summary of measured porosity of specimens.

| | **Porosity [%]** |
|---|---|
| w/b 0.18 | 7.3 |
| w/b 0.20 | 8.98 |
| w/b 0.22 | 6.48 |
| w/b 0.18 vac | 0.67 |
| w/b 0.20 vac | 0.42 |
| w/b 0.22 vac | 1.58 |

In the figure 7, the effect of porosity on the mechanical strengths of samples is presented. The principle is similar to that described by Malecot et all [15]. This figure is a summary of results and influence of porosity on mechanical strength is apparent. The workability of the mixture is also different. When the water content was 22%, mixture was too fluid and it was not possible to achieve such a low porosity. Further reduction of the water coefficient under w/b = 0.18 would make homogenization impossible.

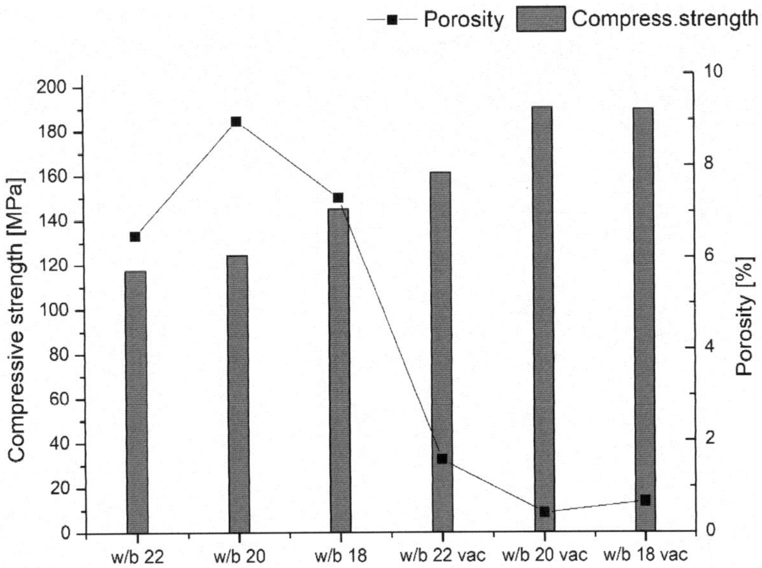


**Figure 7.** Summary of the influence of porosity on compressive strength after 28 days of hydration.

As described in figure 7, reduced porosity increases compressive strength, which is consistent with the results published by Edwin et al. [9]. The effect of mixing under reduced pressure on the resulting strength is apparent.

## 4. Conclusion

The series of samples of RPC mixtures were prepared with a goal to determine the influence of the type of mixing (standard/vacuum) and the water content on the mechanical properties and the total porosity of the designed RPC. The selected w/b ratios varied from 0.18 up to 0.22 with a 0.02 step. Considering the workability and final mechanical parameters, the w/b ratio of 0.2 was determined as optimal for the proposed RPC mixtures. The results indicate, that the large increase in flexural (up to 37%) and compressive (up to 34%) strengths can be achieved by applying vacuum during the mixing process. Significant decrease of porosity of the samples prepared by vacuum mixing was observed. The porosity was reduced from 7.3%, 8.98% and 6.48% to 0.67%, 0.42% and 1.58% in case of samples with w/c = 0.18, 0.20 and 0.22 respectively. It was proven, that applying the vacuum mixing technology in production process can significantly improve the homogeneity and load bearing capacity of the RPC materials, which makes them more suitable for several demanding applications as ballistic protective structures or thin shell structures.

**Acknowledgments**
This outcome has been achieved with the financial support by the project of specific research FCH-S-19-5834.

**References**

[1] Novotný R, Bartoníčková E and Kotrla J 2019 The effect of burnt lime addition on hydration of Ultra-high performance cementitious composites *IOP Conference Series: Materials Science and Engineering* doi:10.1088/1757-899X/583/1/012004
[2] Mohammed H 2015 *M.Sc. thesis, The University of Akron* p 21
[3] Richard P and Cheyrezy M H 1994 *Concrete Technology:Past, Present, and Future S. Francisco* pp 507–518
[4] Coppola L, Troli R, Collepardi S, Borsoi A, Cerulli T and Collepardi M 1996 *L'IndustriaItaliana del Cemento* pp 112–125
[5] Liu S G, Xiang Z, Huang R H, Wang D H and Ju Y Z *IOP Conference Series: Materials Science and Engineering*
[6] Song J and Liu S 2016 Properties of reactive powder concrete and Its application in highway bridge *Advances in Materials Science and Engineering* p 7
[7] Lea's *Chemistry of Cement and Concrete* ISBN 978-0-7506-6256-7
[8] Black L, Dyer D T, Schutter G D and Belie N D *Sustainability of Construction Materials*
[9] Edwin R S, Gruyaert E, Dils J and Belie N D 2017 Influence of Vacuum Mixing on the Carbonation Resistance and Microstructure of Reactive Powder Concrete Containing Secondary Copper Slag as Supplementary Cementitious Material *Procedia Engineering* doi:10.1016/j.proeng.2017.01.367
[10] Dils J, Boel V, Schutter G D and Belie N D 2015 Vacuum mixing technology to improve the mechanical properties of ultra-high performance concrete *Materials and Structures* doi:10.1617/s11527-014-0416-2
[11] Edwin R S, Gruyaert E and Belie N D 2017 Influence of intensive vacuum mixing and heat treatment on compressive strength and microstructure of reactive powder concrete incorporating secondary copper slag as supplementary cementitious material *Construction and Building Materials* doi:10.1016/j.conbuildmat.2017.08.036
[12] Dils J, Boel V, Schutter G D and Belie N D *Materials and Structures*
[13] Richard P and Cheyrezy M 1995 Compostion of reactive powdre concretes *Cement and Concrete Research*
[14] Zdeb T, Boel V, Schutter G D and Belie N D 2019 Effect of vacuum mixing and curing conditions on mechanical properties and porosity of reactive powder concretes *Construction and Building Materials* doi:10.1016/j.conbuildmat.2019.03.116
[15] Malecot Y, Zingg L, Briffaut M and Baroth J 2019 Influence of free water on concrete triaxial behavior: The effect of porosity *Cement and Concrete Research* doi:10.1016/j.cemconres.2019.03.010

# The development of ultra-lightweight concrete based on foam glass aggregate.

**J Zach[1], J Bubenik[1] and M Sedlmajer[1]**

[1] Brno University of Technology, Faculty of Civil Engineering, Department of Building materials and Components, Veveri 95, 602 00 Brno, Czech Republic

Email: zach.j@fce.vutbr.cz

**Abstract**. Foam glass is currently a material produced from waste glass that cannot be used in the glass industry. Therefore, it is currently made from secondary raw materials obtained by recycling unusable waste glass. Due to the large proportion of closed pores in the material structure, the foam glass achieves excellent thermal insulation properties and limited water absorption. The porous structure of the glass is relatively regular, and the pores predominantly have a round character; therefore, the foam glass exhibits very good mechanical properties at a low density. Foam glass in the form of aggregate represents an interesting raw material from which it is possible to produce various types of light composites with a very good ratio of thermal insulation and mechanical properties. The contribution describes the results of research in the field of the development of ultra-lightweight aggregate concretes with subsequent use for producing prefabricated thermal insulating masonry elements for the construction of energy efficient buildings.

## 1. Introduction

Foam glass is a material that was invented in 1930, but it is currently experiencing its renaissance due to its application on the building materials market gradually growing. At the beginning of its creation, foam glass was produced as a cork replacement for military vessels. The glass began to be used in the construction industry around 1960, but due to its high price, it was applied in very small quantities. In the last few years, a high emphasis on waste-free management, recycling, ecology and the environment has created a unique opportunity for this material to break through into construction practice [1, 2].

Foam glass is used in the construction industry in the form of thermal insulation boards or as aggregate (angular and rounded). The production of both forms of the material is very similar. The waste glass is ground into fine flour up to 50 microns, dried and then homogenised together with blowing agents and other additives (most of them are substances reducing the melting point of glass). The prepared raw material mixture is placed either on a belt or in a mould and passed through a tunnel furnace at a temperature of 800 to 900°C, and then either a gradual controlled cooling – the formation of plate foam glass or rapid intensive cooling – the formation of foam glass aggregate [1, 2].

The contribution deals with the use of foam glass aggregate as a lightweight aggregate for producing ultra-lightweight aggregate composites. The objective of the research work performed was to design concrete with the best possible ratio of thermal insulation and mechanical properties, which could be used for mechanically loaded applications, for example, to produce masonry fittings. The design of the element was based on findings obtained within the previous research work [3, 4].

Based on our experience with light-weight concrete, cement class CEM I 42.5R was chosen as a binder. Six test mixtures were designed and mixed to optimise the production formula. The key thermal insulating, physical and mechanical properties were determined on the specimens produced.

The determination of the thermal conductivity in the dependence on humidity was also performed on the selected formula to predict its real behaviour after incorporation into the structure.

Furthermore, the design of the masonry fitting according to standard shapes used to produce the concrete fittings in construction practice was carried out, and the thermal properties according to ČSN EN 1745 were determined.

The properties of the developed concretes were tested in the prototype production of the masonry fittings, i.e. two productions of the prototype fittings were performed, and in both cases, it was necessary to modify the selected formula (No. 1) to produce test specimens. Due to a slightly different way of producing the test specimens in the laboratory and in the manufacturing plant, a higher amount of cement was designed for the formula.

## 2. Production of test specimens

As part of the research on the developed material, six formulas of ultra-lightweight aggregate concrete with aggregate based on foam glass were designed. The objective was to create a compact cementitious composite that will be characterised primarily by a unique ratio of mechanical and thermal insulation properties. All the prepared mixtures were made to optimise the subsequently produced formula to produce the thermal insulating masonry fittings. The test formulas varied in the amount of cement (the minimum amount was 60 kg/m$^3$, and the maximum was 150 kg/m$^3$), and the use of cellulose fibres from recycled paper and their effect on the resulting composite was monitored for the last formula. The individual formulas are described in the following table.

**Table 1.** Test formulas of ultra-lightweight aggregate concrete for 1 m$^3$.

| **Entry component** | | Formula 1 | 2 | 3 | 4 | 5 | 6 |
|---|---|---|---|---|---|---|---|
| | | 1 m$^3$ | | | | | |
| **Cement CEM I 42.5 R** | [kg] | 150 | 130 | 100 | 80 | 60 | 130 |
| **Aggregate 4 - 16 mm REFAGLASS** | [l] | 1000 | 1000 | 1000 | 1000 | 1000 | |
| **Water** | [l] | 57 | 52 | 48 | 44 | 40 | 88 |
| **Plasticizing admixture MURAPLAST FK 19** | 0.8 % from $m_c$ [kg] | 1.2 | 1 | 0.8 | 0.64 | 0.48 | 1 |
| **Fiber recycled paper** | 7.5 % from $m_c$ [kg] | - | - | - | - | - | 9.75 |
| **Water-cement ratio** | [-] | 0.40 | 0.40 | 0.48 | 0.55 | 0.67 | 0.68 |

Two types of test specimens were made from the lightweight aggregate concrete to determine the key properties.

- Cubes 150 x 150 x 150 mm (determination of mechanical and physical-mechanical properties)
- Square specimens 200 x 200 x 35 mm (determination of mechanical and thermal-insulation properties)

**Figure 1.** Made test samples of ultra-light concrete.

**Figure 2.** Detail of lightweight concrete surface.

## 3. Methodology

Several measurements were performed on the specimens produced to determine the required key properties. The measured properties were chosen regarding the subsequent use in practice. Specifically, it was the determination of:

1. The density in a fresh/hardened state (ČSN EN 12350-6, ČSN EN 12390-7) [5, 6],
2. The compressive strength of the concrete (ČSN EN 12390-3) [7],
3. The thermal insulating properties – coefficient of thermal conductivity (ČSN EN 12667, ISO 8301) [8, 9].

### *3.1. Preparation of the test specimens*

To produce the test specimens, moulds provided with a release agent and the necessary raw materials had to be prepared first. Since foam glass is a lightweight aggregate, it is necessary to dose it by volume, i.e. the other raw materials (cement, water and plasticiser) were dosed by weight. The mixture was homogenised in a forced circulation mixer and subsequently placed by means of the vibro-press method into test moulds to produce the test specimens.

## 4. Results and discussion

### *4.1 Determination of mechanical properties*

A consistency test was carried out on the designed formulas using a cone slump test, and the density in the fresh and hardened state was determined. Finally, the compressive strength was determined 28 days after mixing. The results are summarised in the following table.

**Table 2.** Overview of values of consistency and mechanical properties, graph of compressive strength.

| Formula | Settlement of fresh concrete [mm] | $D_{FC}$ [kg/m$^3$] | $D_{HC}$ [kg/m$^3$] (28 days) | Compressive strength $f_{c.28}$ [N/mm$^2$] |
|---|---|---|---|---|
| **1** | 0 | 580 | 550 | 1.54 |
| **2** | 0 | 530 | 500 | 1.42 |
| **3** | 0 | 510 | 460 | 1.25 |
| **4** | 0 | 440 | 410 | 1.11 |
| **5** | 0 | 440 | 390 | 0.88 |
| **6** | 0 | 630 | 560 | 1.52 |

It is apparent from the measured density values that the increase in values occurred with the use of cellulose fibres, and the fibre-free formulas, on the other hand, have lower density values. The lowest value has a formula No. 4 (390 kg/m$^3$), which has only 60 kg of cement and is not fibre reinforced. The difference between the density values in the fresh and hardened state is caused by the loss of mixing water.

Regarding the compressive strength evaluation, the dependence on the density and the amount of cement used is visible. As the amount of cement decreases, the compressive strength decreases relatively linearly. When comparing the formulas with the same amount of cement (Formulas 2 and 6), it can be seen that after the addition of the fibres, the concrete strength slightly increased. This was approximately a 6% strength increase.

### *4.2 Determination of thermal insulation properties*

The characteristic of the lightweight aggregate concrete being developed for use on thermal insulating fittings is the thermal conductivity coefficient. The thermal conductivity was determined by the stationary plate method using a FOX 200 instrument at a temperature gradient of 10 K with a mean temperature of 10°C. The thermal conductivity coefficient λ was determined on specimens of 200 x 200 x 35 mm. The measured values can be seen in the following table.

**Table 3**. Overview of values of coefficient of thermal conductivity in dried state.

| Formula | m [g] | Dimensions [mm] | | | D [kg/m$^3$] | ø D [kg/m$^3$] | λ [W/(m.K)] | | | ø λ [W/(m.K)] |
|---|---|---|---|---|---|---|---|---|---|---|
| | | b | l | h | | | | | | |
| **1** | 2780.2 | 299 | 299 | 63.37 | 490.7 | **500** | 0.1226 | 0.1201 | 0.1193 | **0.1223** |
| | 2848.3 | 300 | 300 | 63.13 | 501.4 | | 0.1254 | 0.1237 | 0.1227 | |
| **2** | 667.3 | 200 | 199 | 35.97 | 466.1 | **460** | 0.1208 | 0.1207 | 0.1207 | **0.1144** |
| | 644.2 | 200 | 200 | 35.99 | 447.6 | | 0.1082 | 0.1081 | 0.108 | |
| **3** | 575.2 | 198 | 200 | 36.42 | 398.9 | **410** | 0.1006 | 0.1004 | 0.1002 | **0.1032** |
| | 601.9 | 200 | 200 | 36.45 | 412.9 | | 0.1061 | 0.1060 | 0.106 | |
| **4** | 558.6 | 200 | 200 | 36.86 | 378.9 | **370** | 0.1022 | 0.1020 | 0.102 | **0.0994** |
| | 528.5 | 199 | 199 | 36.00 | 370.7 | | 0.0969 | 0.0967 | 0.0965 | |
| **5** | 523.9 | 200 | 199 | 35.75 | 368.2 | **360** | 0.0961 | 0.0959 | 0.0958 | **0.0934** |
| | 497.9 | 200 | 200 | 36.09 | 344.9 | | 0.0911 | 0.0909 | 0.0908 | |
| **6** | 730.1 | 200 | 200 | 36.42 | 501.2 | **510** | 0.1131 | 0.1130 | 0.1130 | **0.1170** |
| | 741.8 | 200 | 200 | 36.10 | 513.7 | | 0.121 | 0.1209 | 0.1209 | |

In order to simulate the real conditions during the incorporation of the developed material into the structure, the change of the thermal insulation properties were monitored in dependence on various humidity exposures. More specifically, it was a dried state (i.e. 0% moisture), 30% moisture, 50% moisture and 80% moisture. The dependence of the thermal conductivity coefficient on the humidity of the environment was demonstrated on the board specimens of formula No. 6, which was chosen for the best ratio of the thermal insulation and mechanical properties.

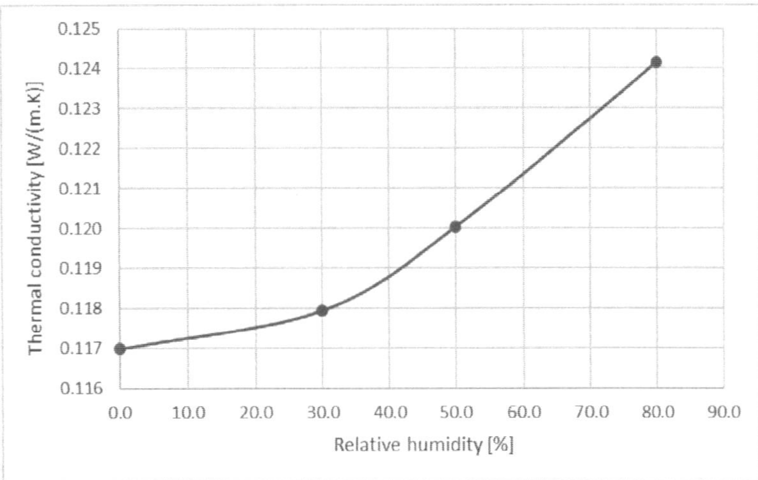


**Figure 3.** Dependence of the thermal conductivity coefficient on moisture of lightweight aggregate concrete, formula No. 6.

The graphical evaluation shows a noticeable increasing tendency of thermal conductivity with increasing environmental humidity. However, the sorption ability of the material being developed is minimal. For the thermal conductivity of the specimen obtained in an environment with 80% humidity, the thermal conductivity coefficient increased only by 6% compared to the values in the dried state. From the measured results, it is evident that the material is only slightly sensitive to humidity change and still retains very good thermal insulation properties. Use in exterior construction applications seems to be appropriate.

## 5. Design of masonry blocks

Based on the measured results and the possibility of cooperation with the producer of concrete goods, the design of the masonry fitting was made according to the shapes used to produce the concrete fittings (dense concrete). The design of the masonry fittings was based on formula 1 mixed in the laboratory.

Experimental production of masonry fittings made of lightweight aggregate concrete for masonry thickness of 200 mm was carried out. The production formula had to be optimized for production (and fibers were finally omitted), but it was based on experience and knowledge from previous research results.

**Table 4.** Optimized production formulas of lightweight concrete for the production of masonry fittings.

| Entry component | | | Formula 1 (1 m³) | Formula 2 (1 m³) |
|---|---|---|---|---|
| **Cement CEM I 42.5 R** | | [kg] | 148 | 158 |
| **Aggregates** | **4–16 mm REFAGLASS** | [l] | 1000 | 1000 |
| **Water** | | [l] | 90 | 90 |
| **Plasticizing admixture REBAMIX 606** | | 0.8 % from $m_c$ [kg] | 1.2 | 1.2 |
| **Water-cement ratio** | | [-] | 0.61 | 0.57 |

The two above-mentioned formulas were produced in the production plant, and the behaviour of the material in the production line was monitored. Because the aggregate of the maximum fraction of 4–8 mm was used in the plant for the production of the existing products, the line was not perfectly adapted for the light aggregate of 4–16 mm fraction, which caused problems, especially in the filling section. To produce perfect functional products, it would be necessary to slightly adjust the filling part of the line and optimise the size of the moulds, especially the wall thickness of the individual products. At the existing thickness, there was a significant collapse of the fresh material (see figures 4 and 5). In consultation with plant workers, there was a positive response when working with material with a significantly lower weight than the existing assortment.

**Figure 4.** Ultra-lightweight masonry blocks - fresh state after pressing.

**Figure 5.** The resulting ultra-lightweight masonry block - after drying.

Test samples were taken from the manufactured fittings (Formula 2) and the following were determined:

- density: 810 kg/m$^3$,
- compressive strength: 3.7 N/mm$^2$.

Due to the small sample size, the thermal conductivity of the fitting concrete was calculated on the basis of measured values on laboratory samples (see table 3). The bulk density of the concrete samples taken from the fitting was significantly higher than that of the test specimens, therefore the measured values (table 3) were approximated by the logarithmic function and the thermal conductivity corresponding to a bulk density of 810 kg/m$^3$ equal to 0.147 W/(m.K). The results are fully comparable with similar masonry elements made of classic lightweight concrete on the construction market [10].

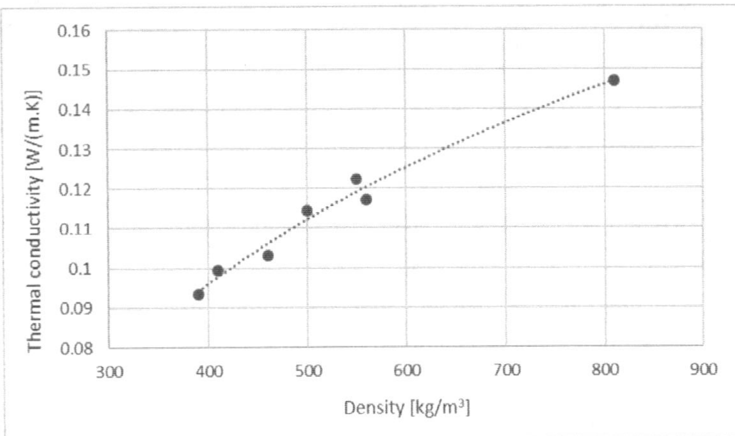


**Figure 6.** Dependence of thermal conductivity on density.

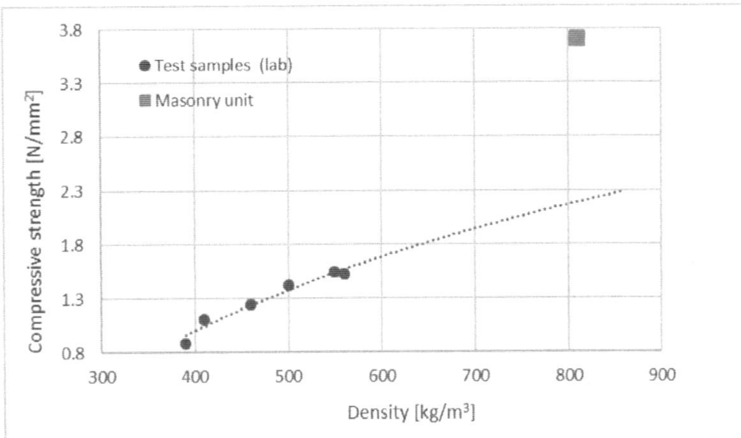


**Figure 7.** Dependence of mechanical properties on density (test samples and masonry unit).

Subsequently, the thermal properties of the proposed fitting under given boundary conditions were determined (see figure 8).

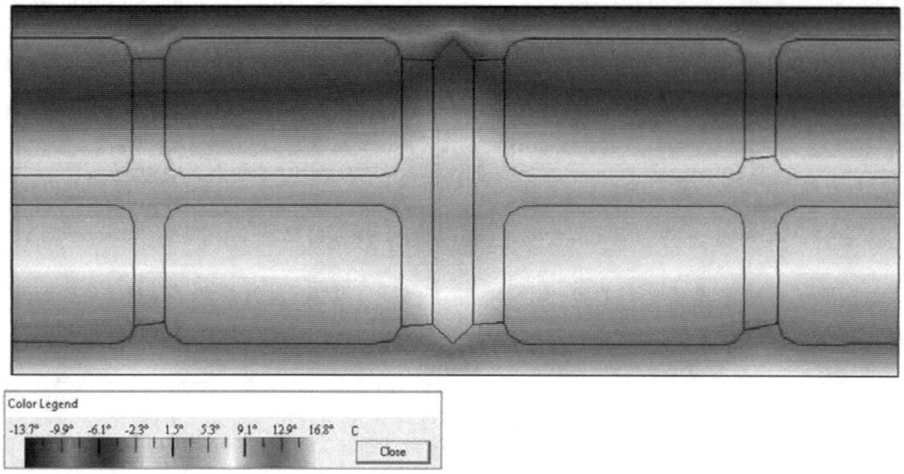


**Figure 8.** Computational distribution of temperature field (IR image) in masonry elements of lightweight aggregate concrete at boundary conditions: $\theta_1 = \theta_i = +21°C$, $\theta_2 = \theta_e = -15°C$.

The heat transfer coefficient of the fitting was calculated to be $U_{unit}$ = 1.03 W/(m$^2$.K) (at a wall thickness of 200 mm).

## 6. Conclusions

Six formulas of ultra-lightweight concrete based on foam glass were designed in laboratory. Developed concretes have a relatively high ratio of mechanical and thermal insulation properties, making this material suitable for many other applications in construction practice. Subsequently, the design and manufacture of the masonry unit was made from a modified laboratory formula of ultra-lightweight concrete. The concrete of the resulting fitting exhibited a higher bulk density (due to the different deposition method compared to the laboratory conditions). However, a significantly higher compressive strength was also achieved, which was 3.7 N/mm$^2$, which at a density of 810 kg/m$^3$ is a very high value, which significantly exceeds the values measured for laboratory samples.

Due to the very good ratio of bulk density, compressive strength and thermal conductivity, the masonry unite exhibits a relatively low heat transfer coefficient at 200 mm thickness and could be used, for example, in internal partition structures between rooms with a temperature difference of up to 10°C. Choosing a suitable internal arrangement of the masonry unite would allow a very low heat transfer coefficient to be achieved at higher masonry thicknesses and it can be argued that the developed lightweight concrete made from foamed glass aggregate has very high potential for wider application especially in civil constructions. Equivalent value of thermal conductivity of masonry block was calculated of $\lambda_{equ}$ = 0.25 W/(m.K). The results are fully comparable with similar masonry elements made of classic lightweight concrete on the construction market. A further improvement could be achieved by changing the geometry of the masonry blocks using air cavities with a smaller thickness perpendicular to the direction of heat flow. When using ultralight concrete for the production of masonry blocks, it would be appropriate to design elements with greater wall thickness of individual elements. Achieving an increase in the thickness of the walls of the blocks would eliminate problems in the production line, increase the load-bearing capacity of the masonry and increase the weight of individual masonry elements.

**Acknowledgements**

This paper was elaborated with the financial support of the project MPO FV20086 "Possibilities of using waste glass powder for the production of lightweight aggregate for subsequent application for the development of modern construction materials with thermal insulating properties".

**References**

[1] Yassaman Y, Schneider A, Baaj H, Tighe S and Youssefi A 2016 *Foam glass lightweight aggregate: The new approach* [online]

[2] Arulrajah, Arul & Disfani, Mahdi & Maghool, Farshid & Udonchai, Artit & Imteaz, Monzur & Du, Yan 2015 Engineering and environmental properties of foamed recycled glass as a lightweight engineering material *Journal of Cleaner Production* **94** 10.1016/j.jclepro.2015.01.080

[3] Zach J, Sedlmajer M, Bubenik J and Drdlova M 2019 Development of lightweight composites based on foam glass aggregate. *IOP Conference Series: Materials Science and Engineering* **583** (1), 012016

[4] Gozzani G, Drdlova M, Prachař V and Popovic M 2016 Characteristics of glass based energy absorbers exposed to the blast load. *Procedia Engineering* **151** pp 214–221

[5] ČSN EN 12350-6 Testing fresh concrete – Part 6: Density (Brusel, Rue de Stassart 36, B-1050) 2019 p 16 (731301)

[6] ČSN EN 12390-7 Testing hardened concrete - Part 7: Density of hardened concrete (Brusel, Rue de Stassart 36, B-1050) 2019 p 16 (731302)

[7] ČSN EN 12390-3 Testing hardened concrete - Part 3: Compressive strength of test specimens (Brusel, Rue de Stassart 36, B-1050) 2019 p 24 (73 1302)

[8] ČSN EN 12667 Thermal performance of building materials and products - Determination of thermal resistance by means of guarded hot plate and heat flow meter methods - Products of high and medium thermal resistance (Brusel, Rue de Stassart 36, B-1050) 2001 p 60 (73 0569)

[9] ISO 8301 Thermal insulation — Determination of steady-state thermal resistance and related properties — Heat flow meter apparatus (Brusel, Rue de Stassart 36, B-1050) 1991 p 30

[10] *Online:* www.archiproducts.com

# Durability of alkali-activated concretes containing cement kiln by-pass dust

**V Bilek Jr.[1,2], V Bilek Sr.[3], L Kalina[1,4], F Khestl[3], J Palovcik[1] and H Simonova[2]**

[1] Brno University of Technology, Faculty of Chemistry, Institute of Materials Science, Purkynova 464/118, 61200 Brno, Czech Republic
[2] Brno University of Technology, Faculty of Civil Engineering, Institute of Structural Mechanics, Veveri 331/95, 602 00 Brno, Czech Republic
[3] Technical University of Ostrava, Faculty of Civil Engineering, Department of Building Materials and Diagnostics of Structures, Ludvika Podeste 1875/17, 708 00 Ostrava-Poruba, Czech Republic
[4] Brno University of Technology, Faculty of Civil Engineering, Institute of Physics, Zizkova 511/17, 602 00 Brno, Czech Republic

Email: bilek@fch.vut.cz

**Abstract**. The aim of the presented work was evaluation of an effect of various conditions on the performance of two developed concretes based on alkali-activated slag and cement kiln by-pass dust (BD). BD was used as a partial replacement of natural aggregates while slag as an aluminosilicate precursor activated by a combination of waterglass and sodium hydroxide solution (resulting silicate modulus of 0.5). The concretes differed only in an activator dose which was 4 and 6% of $Na_2O$ with respect to the slag weight. The prepared specimens were sealed-cured for the first 28 days and then their resistance to freeze-thaw cycles and aggressive solutions (ammonium nitrate, acetic acid and sulphates) was tested. Evolution of dynamic modulus showed that both concretes resisted well to freeze-thaw cycles as well as to most solutions, where their dynamic modulus tended to increase in most cases or did not change significantly. Only the concrete with 4% $Na_2O$ showed poor resistance to acetic acid solution as the specimens completely disintegrated until 50 days.

## 1. Introduction

Nowadays, alkali-activated materials (AAMs) undergo an unprecedented scientific interest. One of the most popular subgroup of AAMs is alkali-activated slag (AAS) due to high early strength achieved even at room temperatures. One of the key parameters for their practical utilization is their durability and resistance to various aggressive conditions. Although wider discussion about the needs for standardization of AAMs and their testing procedures have already started, it is not finished yet. Therefore, durability aspects of AAMs are usually tested on the basis of standards for ordinary Portland cement (OPC) systems and compared to them.

Chemical degradation of the silicate binders can take place via two basic mechanisms and their combinations. These are (i) the formation of new products with deteriorating effect on a binder or a composite structure such as ettringite formed during the sulphate attack and (ii) chemical changes in a binder phase. The main hydration product of AAS is calcium-aluminium-silicate-hydrate (C-A-S-H) [1] from which particularly calcium cations can be leached out that consequently causes polymerization of silicate chains. Calcium cations can also be replaced by other ions.

Typical combination of the above mentioned degradation processes is the action of magnesium sulphate which induces the formation of gypsum but also destabilization of C-A-S-H due to the replacement of $Ca^{2+}$ by $Mg^{2+}$ [2, 3]. Unlike $MgSO_4$, sodium sulphate was reported having much lowereffect on AAS [2, 3] or can be even better than curing in a humid chamber [4].

Decalcification of the C-A-S-H gel can be caused by the action of various organic and inorganic acids as well as acidic salts, typically ammonium nitrate [5, 6]. Very important factor that affects the degradation extent is a solubility of a resulting calcium salt, because the insoluble products can act as a barrier for further degradation [7].

AAS is usually found being more resistant to the above mentioned degradation effects compared to OPC [2, 4, 6, 8, 9]. which is attributed to the different composition of starting materials as well as the nature of hydration products, porosity etc. Also the frost resistance of AAS, key durability parameter particularly in cold and mild climate, was reported being better than that of OPC [7, 10, 11], while AAS is usually more prone to carbonation [12, 13].

Nevertheless, this work is not focused on the comparison of AAS and OPC but to the evaluation of performance of two AAS-based concretes developed within the projects mentioned in the acknowledgement at the end of this paper. These concretes have a potential to be used in practice but before that it is necessary to test them in terms of various durability criteria. Therefore, the selected results are presented in this paper. They can also serve for comparison with other similar materials.

## 2. Experimental part

### *2.1. Materials and concrete composition*

Ground granulated blast furnace slag with the Blaine fineness of 400 $m^2/kg$ was alkali-activated with the combination of waterglasshaving silicate modulus of 1.85 and 50% sodium hydroxide solution. The silicate modulus of the resulting activator mixture was 0.5 and dry matter ($Na_2O + SiO_2$) content of 44.2%.Slag was combined with a natural aggregate, namely sand (0–4 mm) and gravel (4–8 mm) in the portions calculated using Füllergranulometric curve. Total aggregate to slag ratio was 3:1 by weight. Sand was partially replaced by BD whose amount was 10% by the slag weight. To ensure the desired workability 1% (relative to slag weight) of a lignosulphonate-based plasticizer was also added. Water to slag ratio of both concretes was 0.37 including water from the activator.

Chemical composition of both slag and BD is given in table 1. Mineralogical composition of these raw materials was determined using X-ray powder diffraction with Rietveld evaluation method where fluorite ($CaF_2$) was used as an internal standard. The measurement was carried out in Bragg-Brentano configuration in the range of 5–90° 2θ, X-ray tube with copper anode with voltage of 30 kV and electric current of 40 mA was used. It revealed that the slag consisted predominantly of an amorphous phase (~84%), calcite ($CaCO_3$, ~9%) and minor crystalline phases typical for slags (akermanite, melilite, merwinite) while BD consisted mainly of amorphous phase (~44%), free CaO (~16%), sylvite (KCl, ~20%), larnite ($C_2S$, ~10%) and arcanite ($K_2SO_4$, ~9%) and traces of quartz ($SiO_2$). Particle size distribution of both slag and BD determined using laser granulometer (in a dry state) is given in figure 1. It can be seen that both these materials have very similar median (around 10 μm), but BD has much broader distribution of particles which is due toits origin.

**Table 1.** Chemical composition of slag and by-pass dust as determined using X-ray fluorescence.

| | CaO | $SiO_2$ | $Al_2O_3$ | $Fe_2O_3$ | MgO | $Na_2O$ | $K_2O$ | $SO_3$ | $Mn_2O_3$ | $TiO_2$ | $P_2O_5$ | Cl |
|---|---|---|---|---|---|---|---|---|---|---|---|---|
| slag | 39.6 | 38.9 | 8.9 | 0.27 | 8.8 | 0.39 | 0.40 | 1.4 | 0.83 | 0.32 | 0.01 | – |
| BD | 25.9 | 6.5 | 2.5 | 1.6 | 0.48 | 0.44 | 17.1 | 5.2 | – | 0.14 | 0.08 | 10.3 |

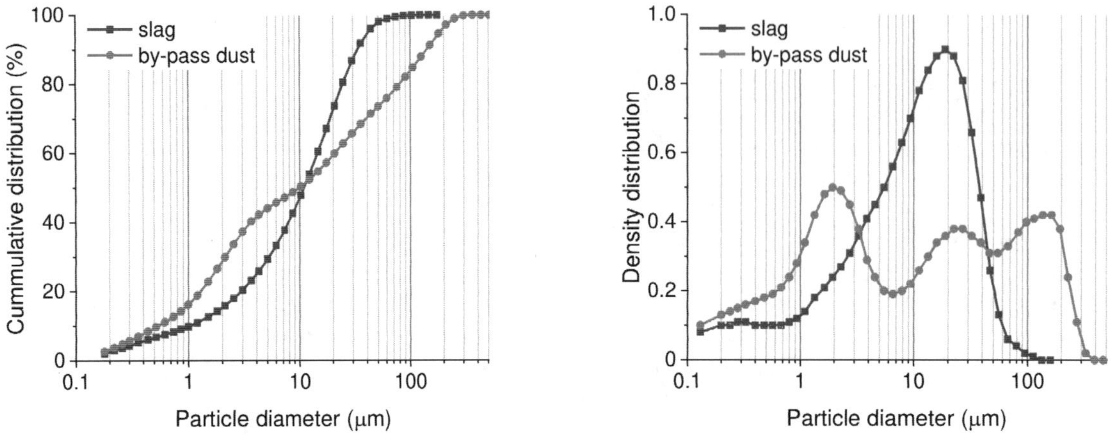


**Figure 1.** Particle size distribution of the slag and cement kiln by-pass dust.

### 2.2. *Specimens preparation and curing regime*

Mixing was performed in a 50l laboratory pan mixer. Firstly, all liquid components were combined together, then slag was added and after that BD and aggregates. A fresh concrete was cast into the moulds to prepare specimens with the dimensions of 40 mm × 40 mm × 160 mm and sealed with a stretch foil. After 24 hours, specimens were demoulded, sealed with stretch foil to prevent moisture loss and stored at laboratory conditions at a temperature of ~25°C until the age of 28 days. After that the specimens were unsealed and transported tovarious degradation media (see section 2.5).

### 2.3. *Strength testing*

During the storage in autogenous conditions, flexural and compressive strengths were tested at the age of 1, 2, 3, 7, 14, 21 and 28 days. At each time, 2–3 specimens were used to test flexural strength while compressive strength was tested on their broken parts.

### 2.4. *Dynamic modulus determination*

Determination of the dynamic modulus $E_d$ was carried out using the device Pundit PL-200 with the frequency of 150 kHz according to equation (1), where the factor $k$ was assumed being 1, $v_p$ was ultrasonic pulse velocity calculated from the time of the passing of the signal and length of the prism and $\rho$ was bulk density of the specimen. Three prisms of each concrete were tested for each degradation conditions (see section 2.5) and each prism was measured three times at each testing time, i.e. just before the start of the degradation and from time to time within the period up to about half of a year.

$$E_d = k \cdot v_p^2 \cdot \rho \quad (1)$$

### 2.5. *Degradation conditions*

Degradation of the specimens took place in basically two ways – by freezing and thawing cycles and in various aggressive solutions. Also water-cured specimens were tested for comparison.

Freezing-thawing cycles were performed according to ČSN 73 1322, i.e. one cycle comprised storage at −20°C for four hours and at 20°C for two hours. Dynamic modulus of the specimens was determined after 68, 125 and 250 cycles.

Aggressive solutions were namely sodium sulphate and magnesium sulphate (concentration of both was 50 g/l of water, i.e. 0.32 and 0.38 mol · dm$^{-3}$, respectively), ammonium nitrate (6 mol · dm$^{-3}$) and acetic acid whose pH was maintained between 3–4 by regular checking using pH meter and additions of a concentrated acid. These tests were carried out at laboratory temperature (~25°C). The concentration of each solution was selected based on a literature review: Sulphate resistance tests across various national standards for Portland cement are summarized in [7], while concentration of ammonium nitrate was selected the same as in [5, 6] used for AAS testing. And finally, acetic acid with pH around 4 was have been already used for the degradation of AAS [9]. The solutions were renewed every month.

## 3. Results

Before the degradation of both concretes started, evolution of their compressive and flexural strength was monitored. The results given in figure 2 showed that increased dose of activator from 4 to 6% $Na_2O$ led to increase in compressive strength which corresponds well with the expectations and also with the previously published results [14]. Such effect was not observed in the case of flexural strength which was similar for both concretes. Although flexural strength values are more scattered, which can also affect the results, the observed trend is related rather to higher internal stresses as a consequence of shrinkage increased by the dose of activator [14, 15]. After 28 days, concretes with 4 and 6% $Na_2O$ reached around 43 and 57 MPa with obvious tendency to further increase.

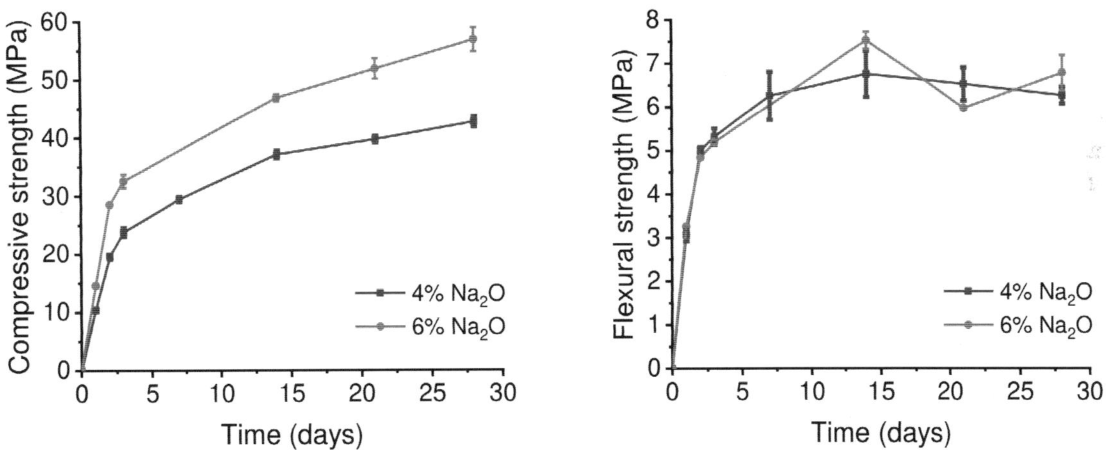


**Figure 2.** Compressive and flexural strength development of concretes in autogenous conditions.

Performance of both concretes in aggressive solutions is expressed as a relative values of dynamic modulus to its initial values obtained just before the degradation is given in figures 3 and 4. Absolute values of dynamic modulus water-cured specimens can be seen later in table 2 showing the results from frost resistance determination and were very similar for all sets of specimens used for the degradation in various solutions, i.e. around 37.0 GPa at the beginning of degradation.

Gradual increase in dynamic modulus of the reference water-cured samples shows the well-known long-term hydration potential of AAS, far beyond 28 days, which was also suggested by the compressive strength development. Approximately the same dynamic modulus evolution in time was observed in the solution of sodium sulphate. Similar trends (in strength evolution) were recorded in [2, 4] where also microstructural analyses were performed but did not show the formation of products related to sulphate attack. Therefore, Sodium sulphate is not an aggressive medium for AAS. This was explained by low availability of calcium and aluminate ions as well as low permeability of AAS. Sodium sulphate can be also used as an activator for AAS and in that case, formation of ettringite, C-A-S-H and layered double hydroxides are formed [16].

Contrary to sodium sulphate, magnesium sulphate has deleterious effect on AAS. According to literature [2, 3], its effect lies in the decalcification and therefore destabilization of C-A-S-H with simultaneous formation of gypsum from leached calcium ions. This results in volume expansion, cracking and thus contribute to AAS degradation. In spite of these findings, it can be surprising that dynamic modulus of both tested concretes did not markedly change in time. Probably, longer exposure is needed to observe more pronounced degradation. However, it is also important to note, that the evaluation of the degradation extent is strongly dependent on the used approach. In this study, the values of dynamic modulus were related to the initial values before the immersion to the solutions, while relating to water-cured specimens is also possible and probably more common. When the latter approach is used, unchanged properties in an aggressive solution and simultaneously their increase in water results in increasing degradation extent. Taking this into account, the results obtained in this work correspond well with the results published in [2]. Also for the case of ammonium nitrate, dynamic modulus did not change noticeably over the testing timescale which is in good agreement with the already published results [6].

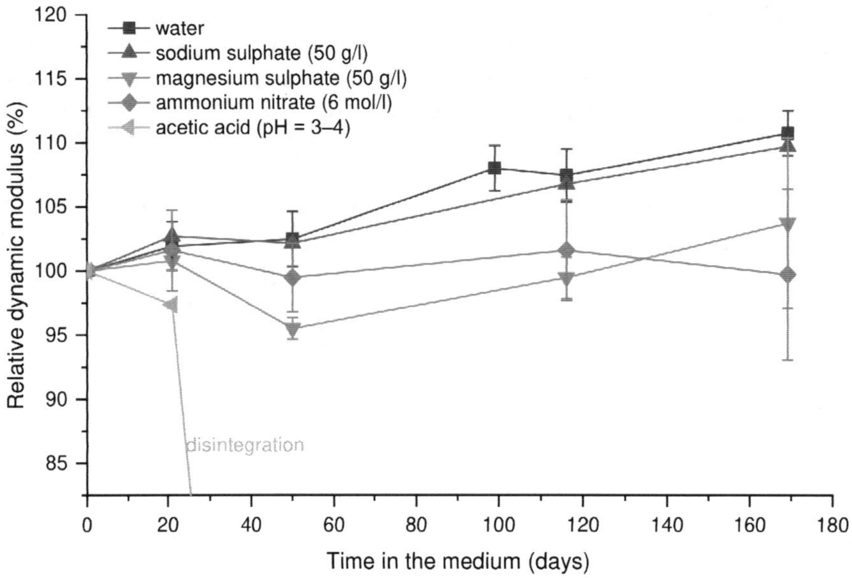


**Figure 3.** Evolution of the dynamic modulus of concrete with 4% $Na_2O$ in various environments.

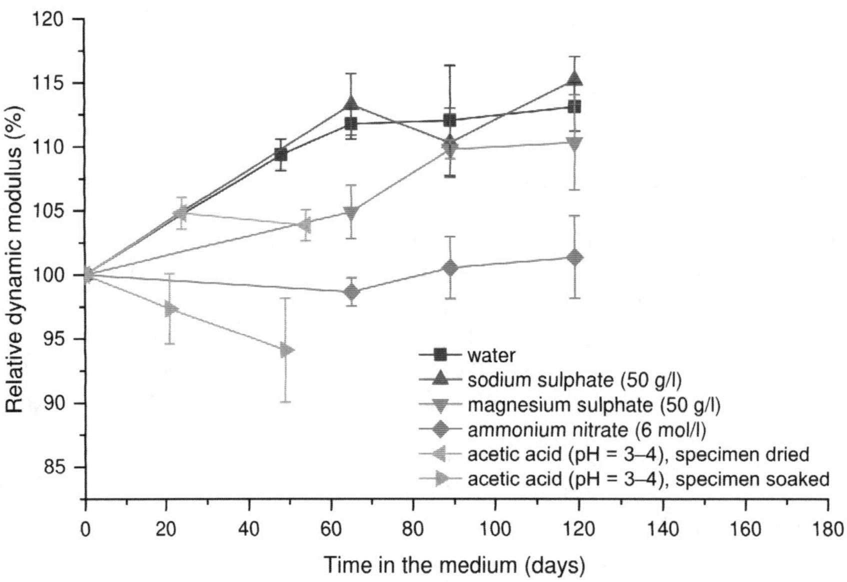


**Figure 4.** Evolution of the dynamic modulus of concrete with 6% $Na_2O$ in various environments.

The last degradation solution was that of acetic acid. Dynamic modulus of concrete with 4% $Na_2O$ only slightly decreased after 21 days but then rapid degradation leading to complete disintegration of the specimens within a few weeks occurred as can be seen from figure 5. Such result was unexpected for us since during some similar tests on AAS the degradation was limited to the first several millimetres from the surface and did not progress deeper even from long-term viewpoint. One possible explanation could be different curing conditions, because in the present study, the specimens were sealed to ensure autogenous conditions, while they were immersed in water in the former case. It was expected that self-desiccated specimens can quickly absorb much higher amount of acid which consequently induces much severe degradation. Therefore, two series of specimens made of later prepared concrete with 6% $Na_2O$ were tested. The first one was immersed in water for five days and then in acetic acid, while another one was left at 20°Cfor five days. In figure 4, they are denoted as soaked and dried, respectively. However, the results (figure 4) showed that none of the two series disintegrated within the first 50 days indicating that there is another reason for the inconsistent performance in acidic environment.

The only difference in composition of both concretes is the activator dose. Better performance of the concrete with the higher activator dose can be related to increased C-A-S-H formation supposed by about one third higher compressive strength compared to concrete with lower activator dose after 28 days. This probably results in lower permeability which reduces the penetration of aggressive media into the specimen. Additionally, it was reported [8] that residual aluminosilicate gel acts as a barrier for further ingress of acid into the specimen. Increased dose of activator and thus higher amount of silicates could also reduce the Ca/Si ratio in the matrix which is also considered contributing to higher stability of C-A-S-H [9].

Both concretes showed very good resistance to freezing and thawing because, up to at least the first 125 cycles (see table 2), the changes in dynamic modulus were only negligible and the above mentioned issue of relating the values arises again. Hopefully, it could be overcome by prolonged curing of the specimens before the start of the degradation as also e.g. ČSN 73 1322 prescribes for blast furnace slag cements, pozzolanic cements etc.

**Figure 5.** Specimens of concrete with 4% $Na_2O$ in acetic acid after 49 days.

**Table 2.** Freezing-thawing resistance of AAS concretes with various dose of activator expressed as changes absolute values of dynamic modulus in GPa ± standard deviation (first line) and in relative values to the start of experiment (second line, in brackets).

| Concrete | Conditions | 0 cycles | 68 cycles | 125 cycles | 250 cycles |
|---|---|---|---|---|---|
| 4% $Na_2O$ | Water (reference) | 36.3 ± 0.8 (100%) | 37.0 ± 0.7 (102%) | 37.2 ± 0.8 (102%) | 40.2 ± 0.7 (111%) |
| | Frozen-thawed | 37.0 ± 0.4 (100%) | 35.0 ± 0.5 (94.6%) | 36.1 ± 0.5* (97.6%) | 30.0 ± 2.8 (81.1%) |
| 6% $Na_2O$ | Water (reference) | 37.4 ± 0.6 (100%) | – | 41.8 ± 0.5 (112%) | 42.3 ± 0.8 (113%) |
| | Frozen-thawed | 37.0 ± 0.6 (100%) | – | 36.2 ± 0.4 (97.8%) | 23.9 ± 0.7 (64.6%) |

*Between the first 125 cycles and the next 125, specimens were immersed in water for 49 days without freezing and thawing. Nevertheless, their dynamic modulus did not change during this period as it was 36.3 ± 0.6 GPa just before the start of the last 125 cycles.

## 4. Conclusions

In this study, resistance of two AAS-based concretes to freezing and thawing cycles and to various aggressive solutions was evaluated in terms of their dynamic modulus development. Both concretes showed very good performance in most of these aspects. Their dynamic modulus started to noticeably decrease after around 250 freezing and thawing cycles. Both concretes behaved similarly also in most solutions showing unchanged or even increasing dynamic modulus compared to that at the beginning of degradation and therefore good resistance to these solutions. Only acetic acid had negative effect on concretes, particularly on that with 4% $Na_2O$, which completely disintegrated during the first 50 days of exposure.

## Acknowledgement

This outcome has been achieved with financial support of the Czech Science Foundation under the project No 18-12289Y and is also a part of a project that has received funding from the European Union's Horizon 2020 research and innovation programme under the Marie Skłodowska-Curie grant agreement No 734833.

The authors also thank to Zuzana Hantakova and Ivan Rachler from chemical laboratory of Horne Srnie cement plant (Cemmac, a.s.) for the facilitation of the slag and BD chemical composition determination.

## References

[1] Puertas F, Palacios M, Manzano H, Dolado J S, Rico A and Rodríguez J 2011 A model for the C-A-S-H gel formed in alkali-activated slag cements *J. Eur. Ceram. Soc.* **31** pp 2043–56
[2] Bakharev T, Sanjayan J G and Cheng Y-B 2002 Sulfate attack on alkali-activated slag concrete *Cem. Concr. Res.* **32** pp 211–6
[3] Ismail I, Bernal S A, Provis J L, Hamdan S and van Deventer J S J 2012 Microstructural changes in alkali activated fly ash/slag geopolymers with sulfate exposure *Mater. Struct.* **46** pp 361–73
[4] Komljenović M, Baščarević Z, Marjanović N and Nikolić V 2013 External sulfate attack on alkali-activated slag *Constr. Build. Mater.* **49** pp 31–9
[5] Varga C, Alonso M M, Mejía de Gutierrez R, Mejía J and Puertas F 2014 Decalcification of alkali-activated slag pastes. Effect of the chemical composition of the slag *Mater. Struct.* **48** pp 541–55

[6] Komljenovic M M, Bascarevic Z, Marjanovic N and Nikolic V 2012 Decalcification resistance of alkali-activated slag *J. Hazard. Mater.* **233-234** pp 112–21
[7] Provis J L and van Deventer J S J 2014 *Alkali activated materials* (Dordrecht: Springer)
[8] Bernal S A, Rodríguez E D, Mejía de Gutiérrez R and Provis J L 2012 Performance of alkali-activated slag mortars exposed to acids *J. Sustain. Cem.-Based Mater.* **1** pp 138–51
[9] Bakharev T, Sanjayan J G and Cheng Y B 2003 Resistance of alkali-activated slag concrete to acid attack *Cem. Concr. Res.* **33** pp 1607–11
[10] Shi C, Krivenko P V and Della R 2006 *Alkali-Activated Cements and Concretes* (Oxon: CRC Press)
[11] Pacheco-Torgal F, Abdollahnejad Z, Camões A F, Jamshidi M and Ding Y 2012 Durability of alkali-activated binders: A clear advantage over Portland cement or an unproven issue? *Constr. Build. Mater.* **30** pp 400–5
[12] Puertas F, Palacios M and Vázquez T 2005 Carbonation process of alkali-activated slag mortars *J. Mater. Sci.* **41** pp 3071–82
[13] Bakharev T, Sanjayan J G and Cheng Y-B 2001 Resistance of alkali-activated slag concrete to carbonation *Cem. Concr. Res.* **31** pp 1277–83
[14] Aydın S and Baradan B 2014 Effect of activator type and content on properties of alkali-activated slag mortars *Compos. Part B-Eng.* **57** pp 166–72
[15] Melo Neto A A, Cincotto M A and Repette W 2008 Drying and autogenous shrinkage of pastes and mortars with activated slag cement *Cem. Concr. Res.* **38** pp 565–74
[16] Mobasher N, Bernal S A and Provis J L 2016 Structural evolution of an alkali sulfate activated slag cement *J. Nucl. Mater.* **468** pp 97–104

IOP Conf. Series: Materials Science and Engineering 1039 (2021) 012013 doi:10.1088/1757-899X/1039/1/012013

# The influence of alkaline activator on immobilization of metals in alkali-activated blast furnace slag

**E Bystrianska**[1] **and J Koplik**[1]

[1] Brno University of Technology, Faculty of Chemistry, Purkyňova 118, 612 00 Brno, Czech Republic

Email: emilia.bystrianska@vut.cz

**Abstract**. In this work the influence of alkaline activator on immobilization of lead and copper in alkali-activated blast furnace slag was investigated. A total of five alkali activators were used; sodium waterglass, potassium waterglass, sodium hydroxide, potassium hydroxide and sodium carbonate. The leaching test according to ČSN EN 12457–4 was used to evaluate the level of immobilization of heavy metals, the leached solutions were analyzed by ICP-OES. For a better understanding of immobilization process, the selected samples were characterized by analytical methods (SEM, XPS). It was found that the degree of immobilization of $Pb^{2+}$ and $Cu^{2+}$ in AAS was very high regardless of the type of alkaline activator used. The high degree of immobilization is caused by formation of insoluble or poorly soluble lead and copper compounds in an alkaline environment while the insoluble compounds have been confirmed by SEM and XPS.

## 1. Introduction

Solidification/stabilization (S/S) is one of the most important techniques for treating wastes containing heavy metal contaminants. The purpose of S/S technology is physically as well as chemically fix heavy metals (or polluted waste) in the solid matrix to reduce their mobility and therefore minimize the threat to the environment. This technology is widely used to dispose of low-level radioactive waste, hazardous and mixed waste. Ordinary Portland cement (OPC) has been the most common solidification material because of its availability and low cost. However, OPC binders are not very effective in stabilizing some heavy metals [1–2].

The efficiency of the heavy metal immobilization is strongly related to the microstructure of hardened paste and total porosity. Hardened alkali-activated slag (AAS) binders have less porosity than OPC binder. Therefore, the mitigation of solution into the AAS binder is limited which results in a lower leachability of toxic waste. Except for total porosity of binders, pH values of pastes play an important role in the process of S/S and type of C-A-S-H gel forming in paste too. In a high alkaline environment, most of the heavy metals precipitate into less soluble or insoluble compounds. Besides low permeability and alkaline pore solution, AAS binders present other important properties including high mechanical strength and high resistance to acid or chloride attack. Due to all these physical and chemical advantages, alkali-activated slag binders are considered to be more effective for waste immobilization than OPC [3–5].

The ability of alkali-activated slag to immobilize heavy metals or other hazardous materials in its structure has been investigated in various studies [6–8]. The efficiency of heavy metal immobilization in alkali-activated slag binders depends on many factors including nature and concentration of alkaline activator used or type of immobilized material and its dosage. Heavy metals addition ($Zn^{2+}$, $Pb^{2+}$, $Cd^{2+}$ a $Cr^{6+}$) up to 2% of slag weight can be well stabilized in NaOH, $Na_2CO_3$, and sodium silicate-activated slag binder [9]. Solidification/stabilization of $Pb^{2+}$ and $Cu^{2+}$ were mostly investigated in fly ash-based geopolymers with high immobilization efficiency [10–13].

The aim of this work is to compare the ability of AAS binders activated by different types of alkaline activators to immobilize Pb and Cu in their structure. In the present work, we have also tried to determine if the heavy metals are stabilized by chemical converting into their less soluble or insoluble forms.

## 2. Experimental procedure

### 2.1. Materials

*2.1.1. Ground granulated blast furnace slag.* Ground granulated blast furnace slag (GGBFS) was obtained from Kotouč Štramberk, Ltd. having a Blaine specific surface area of 400 $m^2 \cdot kg^{-1}$. The GBFS was mostly amorphous but it also contained crystalline phases such as åkermanite, calcite, merwinite, and quartz (measured by X-ray powder diffraction). The chemical composition of GGBFS measured by the X-ray fluorescence technique used for alkali activation is given in table 1.

**Table 1**. Chemical composition of ground granulated blast furnace slag.

| GGBFS Kotouč Štramberk [wt. %] | | | | | | | | | |
|---|---|---|---|---|---|---|---|---|---|
| CaO | $SiO_2$ | MgO | $Al_2O_3$ | $SO_3$ | $TiO_2$ | $K_2O$ | MnO | $Na_2O$ | $Fe_2O_3$ |
| 41.1 | 34.7 | 10.5 | 9.1 | 1.4 | 1.0 | 0.9 | 0.6 | 0.4 | 0.3 |

*2.1.2. Alkaline activators.* For this work five alkaline activators were used: Liquid sodium waterglass (S-WG) – Silicate modulus $M_s = 1.96$; Liquid potassium waterglass (P-WG) – Silicate modulus $M_s = 1.67$; Sodium hydroxide (SH); Potassium hydroxide (PH) and Sodium carbonate (SC). Sodium and potassium hydroxide were used as 50% solution. Sodium carbonate was dissolved in water and cooled to ambient temperature before alkali activation. The $Na_2O/K_2O$ dosage of the alkaline activator was kept the same.

*2.1.3. Heavy metals.* The heavy metal-containing compounds $Pb(NO_3)_2$ p.a. and $Cu(NO_3)_2.3H_2O$ p.a. have been used in sample preparation. The amounts of heavy metals ($Pb^{2+}$, $Cu^{2+}$) adding to the hydrating samples were on level 1% of slag weight.

### 2.2. Tests and methods

*2.2.1. Sample preparation.* The pastes were prepared at the water to binder ratio equal to 0.38 (w/b was kept the same for each activator used). The concentration of alkaline activator in the slag binder was 6% of $Na_2O/K_2O$ of the slag content for each activator used. The alkaline activators were fully dissolved in water and after that mixed with slag. The pastes were mixed in a standard laboratory mixer for 180 s, cast into moulds (20 × 20 × 100 mm) and vibrated for 30 s to remove large air bubbles. All the samples were matured in the moulds for the first 24 h at 20 C and 100% relative humidity. After that samples were open-air cured at ambient conditions.

*2.2.2. Mechanical properties testing.* Specimens with dimensions 20 × 20 × 100 mm were used for compressive strength determination using the Desttest 4310 Compact A (Beton System, Ltd.). The compressive strength of the alkali-activated slag was measured after 1, 7, and 28 days on three samples.

*2.2.3. Leaching test.* The leaching tests were performed using an adaptation of standard ČSN EN 12457–4:2002. After 28 days of curing, specimens of dimensions 20 × 20 × 100 mm were placed into the PET bottle with distilled water at a liquid/solid ratio 10:1 and rotated at 10 rpm for 24 h.

The leachates were analysed on ICP-OES spectroscope (Inductively coupled plasma optical emission spectrometry).

*2.2.4. Characterization of materials, hydration products and microstructure analysis.* The GGBFS was characterized by XRD and XRF methods. XRD measurement was performed using a PANanalytical Empyrean X-ray diffractometer. The tube voltage and current were 40 kV and 30 mA, respectively. The tube anode was CuK$\alpha_1$ (1.54 Å). The sample was step-scanned from 5° to 90° 2θ using vertical high-resolution goniometer with a step size of 0.01313 2θ. Times per step were 96 s.

The chemical composition of GGBFS was determined with the Xenemtrics EX-6600 XRF spectrometer. Measurement of GGBS powder was performed in vacuum using accelerated voltage 30 kV.

Hydration products and microstructure investigations of AAS with and without heavy metals addition were carried out by SEM and XPS (X-ray photoelectron spectroscopy). These investigations were performed on samples after 28 days of curing.

The SEM analysis was performed on ZEISS EVO LS 10 with the energy dispersive X-ray spectroscopy (EDS) detector. Accelerated voltage was set to 15 kV and the working distance was 12 mm. All samples were sputtered by gold to obtain good surface conductivity.

The XPS analyses were conducted with Kratos Axis Ultra DLD spectrometer using a monochromatic Al Kα (hν = 1486.7 eV) X-ray source operating at 150 W (10 mA, 15 kV). The high-resolution spectra were measured with a step size of 0.1 eV and 20 eV pass energy. The instrument base pressure was $2 \times 10^{-8}$ Pa.

## 3. Results and discussion

### *3.1. Compressive strength*

As can be seen from data obtained (table 2) the type of the alkaline activator used strongly influenced the compressive strength of AAS pastes. Without heavy metals addition the highest 28-day compressive strengths were obtained from AAS samples activated by sodium and potassium waterglass. Effect of cations ($Na^+$,$K^+$) from silicate activators cannot be described because of different silicate modulus of waterglass activators used. However, in sodium and potassium hydroxide activation with same dosage, KOH-activated samples formed denser microstructure [14]. Sodium carbonate-activated slag pastes required a prolonged setting time so they were kept in moulds for 3 days. The longer hardening process is related to the slow development of the alkalinity required for initializing slag dissolution. Although, after 28 days of curing, the sodium carbonate-activated samples gained relatively high compressive strengths (56.9 MPa) [15].

The influence of 1% $Pb^{2+}$ addition on the compressive strength of alkali-activated slag is not very significant. In waterglass-activated samples the measured 1-, 7- and 28-day compressive strengths were very comparable to the reference samples. Samples with $Pb^{2+}$ addition activated by sodium or potassium hydroxide exhibited strength loss up to 30% after 28 days of curing. The addition of 1% $Cu^{2+}$ influenced the compressive strength more significantly regardless of alkaline activator used. Sodium carbonate-activated samples were most affected by $Cu^{2+}$ addition and the 28-day compressive strengths reached only half the strength compared to the reference samples.

The nature of alkaline activator plays an important role in compressive strength development. The highest compressive strengths were obtained by sodium and potassium waterglass and the lowest compressive strength were measured on the samples activated by sodium or potassium hydroxide. Similar results were found in other studies [9, 16].

**Table 2.** Compressive strength of AAS samples with different alkali activator used.

| | Compressive strength [MPa] | | |
|---|---|---|---|
| Sample | 1 day | 7 days | 28 days |
| Na-WG | 19.4 | 69.2 | 102.1 |
| Na-WG (1% Pb) | 18.6 | 68.0 | 94.9 |
| Na-WG (1% Cu) | 11.2 | 39.6 | 77.5 |
| NH | 10.4 | 22.0 | 37.4 |
| NH (1% Pb) | 4.6 | 16.3 | 24.6 |
| NH (1% Cu) | 4.3 | 9.3 | 22.6 |
| K-WG | 20.6* | 50.6 | 76.5 |
| K-WG (1% Pb) | 20.1* | 44.9 | 78.1 |
| K-WG (1% Cu) | 10.5* | 17.0 | 59.4 |
| KH | 10.2 | 30.2 | 44.3 |
| KH (1% Pb) | 5.2 | 19.1 | 36.3 |
| KH (1% Cu) | 4.4 | 12.7 | 31.5 |
| NC | - | 51.1 | 56.9 |
| NC (1% Pb) | - | 13.1 | 45.8 |
| NC (1% Cu) | - | 10.1 | 28.4 |

*samples measured after 3 days of curing

### *3.2. Leaching tests*

Leaching tests were performed based on the ČSN EN 12457–4:2002. Three specimens from each sample were subjected to a leaching test in distilled water for 24 hours. Results from heavy metal leaching are given in table 3 and table 4. The results showed that the immobilization efficiency of both heavy metals (Cu and Pb) is very high. The highest Pb release occurred in sodium waterglass-activated samples which could be a result of microcracks observed in these samples. The leaching rate of $Pb^{2+}$ corresponds to the findings obtained by Deja et al who reported that amounts of $Pb^{2+}$ ions leached from waterglass-activated mortars are higher than from $Na_2CO_3$-activated mortars [9].

Samples with $Cu^{2+}$ addition also showed a low concentration of heavy metal in leachates. From the data obtained the best copper and lead stabilization is achieved by potassium waterglass-activated slag binder. It should be noted that concentrations of both heavy metals in all samples are low and all types of alkaline activators provide safe heavy metals immobilization. High immobilization efficiency of $Pb^{2+}$ and $Cu^{2+}$ in alkali-activated GGBFS was presented in the work of Giergiczny et al [12].

**Table 3.** Concentration of $Pb^{2+}$ in leachate.

| sample | Na-WG (1 % Pb) | NH (1 % Pb) | K-WG (1 % Pb) | KH (1 % Pb) | NC (1 % Pb) |
|---|---|---|---|---|---|
| c($Pb^{2+}$) [mg/l] | 3.070 ±1.146 | 0.870 ±0.345 | 0.413 ±0.039 | 2.797 ±0.050 | 1.240 ±0.388 |

**Table 4.** Concentration of $Cu^{2+}$ in leachate.

| sample | Na-WG (1 % Cu) | NH (1 % Cu) | K-WG (1 % Cu) | KH (1 % Cu) | NC (1 % Cu) |
|---|---|---|---|---|---|
| c($Cu^{2+}$) [mg/l] | 0.950 ±0.138 | 0.930 ±0.043 | 0.443 ±0.072 | 0.980 ±0.024 | 0.470 ±0.076 |

### *3.3. Microstructure of pastes*

The microstructure of hardened alkali-activated pastes with heavy metals addition was investigated by SEM analysis. All the AAS samples (regardless of alkaline activator used) had a very dense structure with C-(A)-S-H gel as the main component. The significant differences between the samples doped with heavy metals and reference samples were not observed. The samples with sodium waterglass activation showed higher amounts of microcracks than any other samples (figure 1). Heavy metal compounds (lead and copper) were rather cumulated in some regions than dispersed through the structure. The precipitated lead compound can be seen in figure2 and hollow compound containing copper is shown in figure 3. From EDS analysis taken in areas with high heavy metal concentration (table 5, table 6) can be seen that lead and copper are primary cumulated with oxygen.

**Table 5.** EDS analysis of the area with high Pb concentration from figure 2.

| Element | O | Na | Mg | Al | Si | Ca | Pb |
|---|---|---|---|---|---|---|---|
| Atom. % | 41.21 | 3.29 | 0.37 | 0.33 | 7.35 | 1.13 | 15.73 |

**Table 6.** EDS analysis of the hollow particle containing copper from figure 3.

| Element | O | Na | Si | Cl | Ca | Cu |
|---|---|---|---|---|---|---|
| Atom. % | 61.89 | 9.54 | 0.72 | 5.31 | 1.22 | 20.34 |

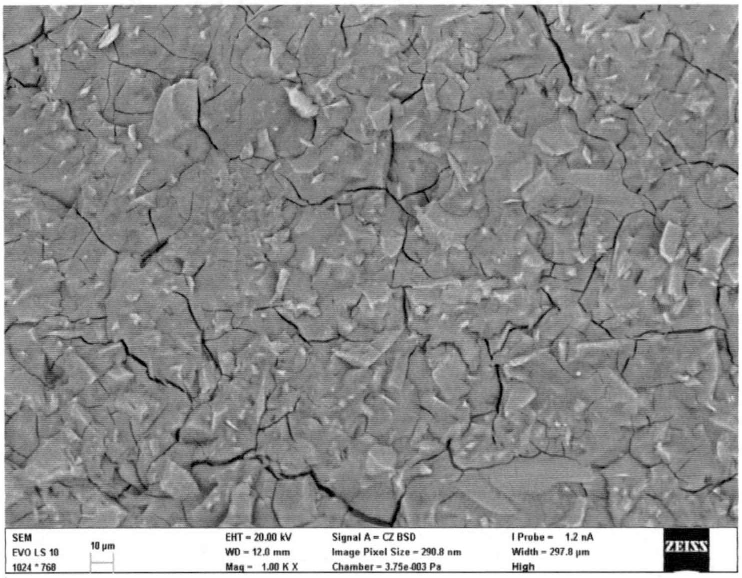


**Figure 1.** Sodium waterglass AAS binder doped with $Pb^{2+}$.

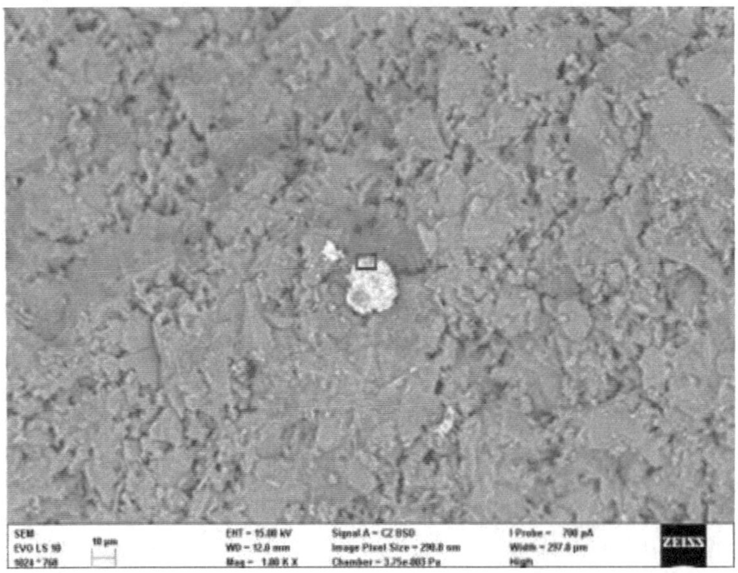


**Figure 2.** Sodium carbonate AAS binder doped with $Pb^{2+}$.

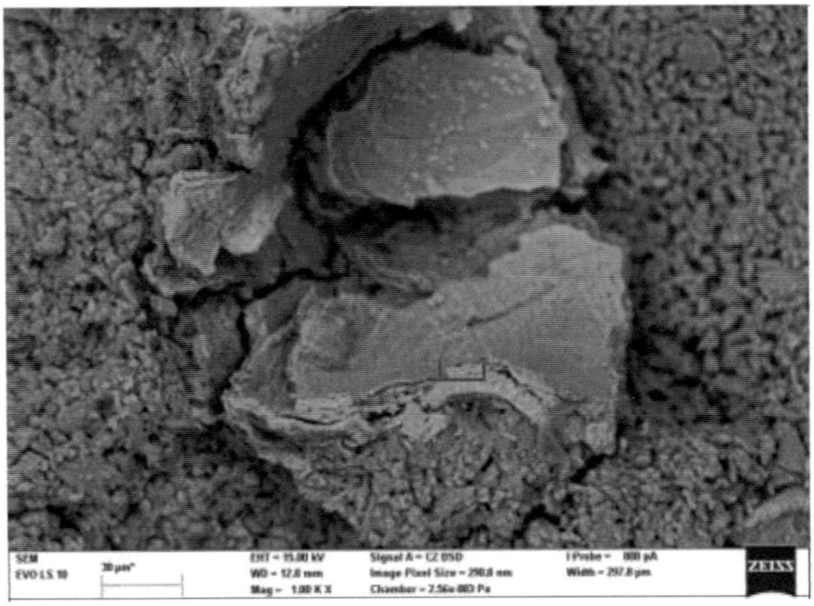


**Figure 3.** Sodium hydroxide AAS binder doped with $Cu^{2+}$.

### 3.4. *XPS analysis*

The XPS spectra of Pb of sample Na-WG (1% Pb) is showed in figure 4. The binding energies of Pb $4f_{7/2}$ and Pb $4f_{5/2}$ are 138.6 eV and 143.3 eV. These values are typical for $Pb(OH)_2$. The highly soluble primary phase $Pb(NO_3)_2$ dissolves during sample preparation and $Pb^{2+}$ ions in an alkaline solution precipitate into an insoluble hydroxide.

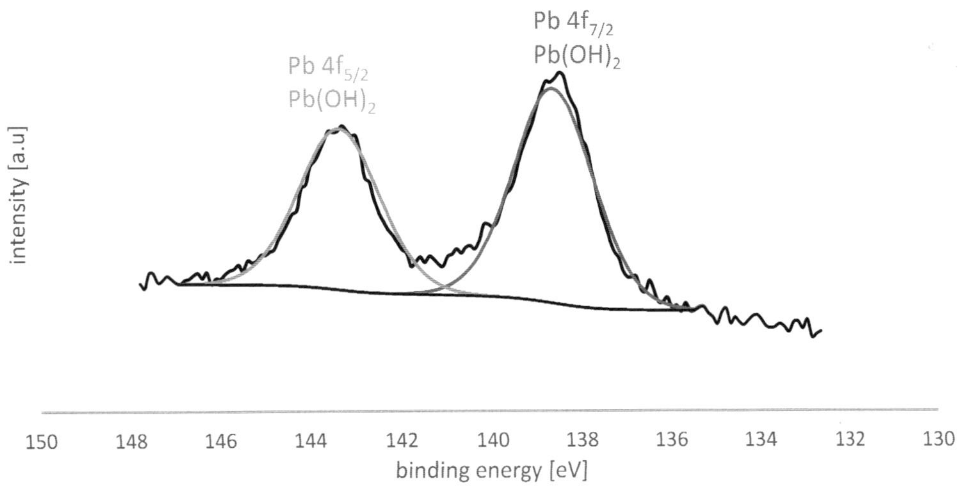


**Figure 4.** XPS spectra of Pb in sample activated by sodium waterglass.

In XPS spectra (figure 5) of a sample containing copper and activated by 50% solution of sodium hydroxide (1% Cu) are shown two different chemical states. The first, Cu $2p_{3/2}$ component with binding energy 935.35 eV with strong shake-up satellites belongs to $Cu(OH)_2$. Shake-up satellites occur when an outgoing electron interacts with a valence electron and excites it (shakes it up) to a higher energy level. As a consequence, the energy core electron is reduced and a satellite structure appears a few eV below (KE scale) the core level position. The second peak, Cu $2p_{3/2}$ with binding energy 933.50 eV, represents the bond between Cu and O. That can be interpreted as the formation of copper oxide or the formation of the bond with the matrix. Both copper compounds (CuO and $Cu(OH)_2$) are insoluble in distilled water [17, 18].

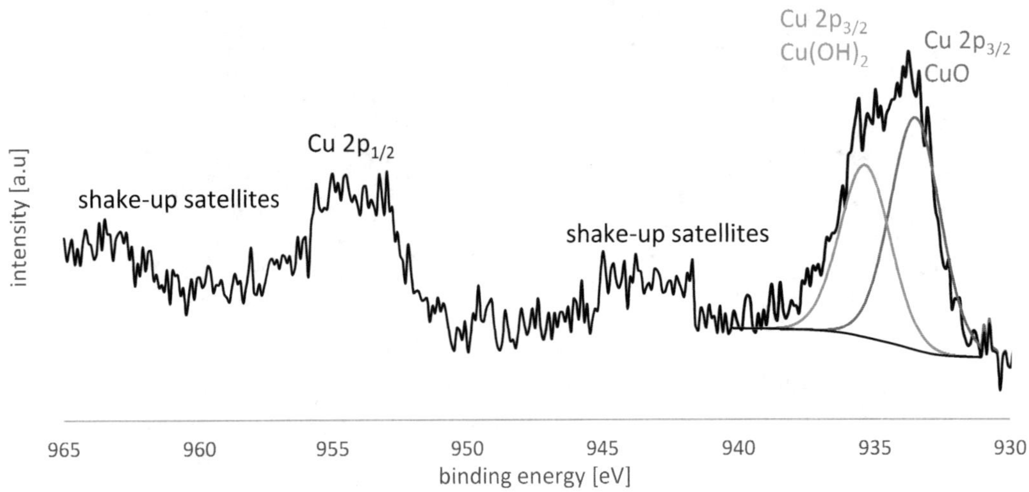


**Figure 5.** XPS spectra of Cu in sample activated by sodium hydroxide.

## 4. Conclusion

The influence of alkaline activator type on heavy metals (Pb and Cu) immobilization was investigated. Five alkaline activators were used: sodium waterglass, potassium waterglass, sodium hydroxide, potassium hydroxide and sodium carbonate. From the results obtained by leaching tests (ČSN EN 12457–4:2002) can be seen that the immobilization efficiency of heavy metals ($Pb^{2+}$ and $Cu^{2+}$) is very high for all types of alkaline activator used. Slightly higher concentrations of heavy metals in leachates were observed when the samples were activated by sodium waterglass. It was probably caused by presence of microcracks which allowed leaching out heavy metals from samples. From SEM observation can be seen that both heavy metals – lead and copper, were rather cumulated in some areas than disperse through the whole structure. Finally, XPS analysis confirmed that Pb and Cu formed insoluble salts in alkali-activated slag binders.

## References

[1] Yilma O, Ünlü K and Cokca E 2003 Solidification/Stabilization of Hazardous Wastes Containing Metals and Organic Contaminants *Journal of Environmental Engineering* **129**(4) pp 366–376

[2] Al-kindi G, Provis J L, Myers R J, San Nicolas R and Van Deventer J S J 2019 Evaluation the Solidification/Stabilization of Heavy Metals by Portland Cement: Alternative matrices for the immobilisation of hazardous wastes *Journal of Ecological Engineering* **20**(3) pp 91–100

[3] Shi C, Fernández-Jiménez A, Rekha P, Bindhu V H and Anjaneyulu Y 2006 Stabilization/solidification of hazardous and radioactive wastes with alkali-activated cements *Journal of Hazardous Materials* **137**(3) pp 1656–1663

[4] Provis J L and Van Deventer J S J 2014 Alkali Activated Materials: State-of-the-Art Report, RILEM TC 224-AAM.

[5] Palomo A, López Dela Fuente J I, Rekha P, Bindhu V H and Anjaneyulu Y 2003 Alkali-activated cementitous materials: Alternative matrices for the immobilisation of hazardous wastes *Cement and Concrete Research* **33**(2) pp 281–288

[6] Qian G, Delai Sun D, Hwa Tay J and Sun G 2003 Characterization of mercury- and zinc-doped alkali-activated slag matrix *Cement and Concrete Research* **33**(8) pp 1251–1256

[7] Shi C, Fernández-Jiménez A, Zhao M, Yang K, Shen R and Zheng Y 2006 Stabilization/solidification of hazardous and radioactive wastes with alkali-activated cements

*Journal of Hazardous Materials* **137**(3) pp 1656–1663
[8] Zhang M, Yang C, Zhao M, Yang K, Shen R and Zheng Y 2017 Immobilization potential of Cr(VI) in sodium hydroxide activated slag pastes *Journal of Hazardous Materials* **321**(8) pp 281–289
[9] Deja J, Sun W, She W and Sun G 2002 Immobilization of Cr6 , Cd2 , Zn2 and Pb2 in alkali-activated slag binders *Cement and Concrete Research* **32**(12) pp 1971–1979
[10] Zhang Y, Sun W, She W and Sun G 2009 Synthesis and heavy metal immobilization behaviors of fly ash based gepolymer *Journal of Wuhan University of Technology-Mater. Sci. Ed.* **24**(5) pp 819–825
[11] Wang Y, Han F, Mu J and Sun G 2018 Solidification/stabilization mechanism of Pb(II), Cd(II), Mn(II) and Cr(III) in fly ash based geopolymers *Construction and Building Materials* **160**(8) pp 818–827
[12] Giergiczny Z, Król A, Zhao M, Yang K, Shen R and Zheng Y 2008 Immobilization of heavy metals (Pb, Cu, Cr, Zn, Cd, Mn) in the mineral additions containing concrete composites *Journal of Hazardous Materials* **160**(2–3) pp 247–255
[13] Zheng L, Wang W, Qiao W, Shi Y, Liu X and Zheng Y 2015 Immobilization of Cu2 , Zn2 , Pb2 and Cd2 during geopolymerization *Journal of Hazardous Materials* **9**(4) pp 642–648
[14] Tänzer R, Jin Y and Stephan D 2017 Alkali activated slag binder: effect of cations from silicate activators *Materials and Structures* **50**(1)
[15] Bernal S A., Provis J L, Myers R Y, San Nicolas R and Van Deventer J S J 2015 Role of carbonates in the chemical evolution of sodium carbonate-activated slag binders: Alternative matrices for the immobilisation of hazardous wastes *Materials and Structures* **48**(3) pp 517–529
[16] Fernandez-Jimenez A, Palomo J and Puertas F 1999 Alkali-activated slag mortars-Mechanical strength behaviour
[17] Watts J F and Wolstenholme J 2005 An Introduction to Surface Analysis by XPS and AES (John Wiley & Sons)
[18] Biesinger M C, Provis J L, Myers R Y, San Nicolas R and Van Deventer J S J 2017 Advanced analysis of copper X-ray photoelectron spectra: Alternative matrices for the immobilisation of hazardous wastes *Surface and Interface Analysis* **49**(13) pp 1325–1334

IOP Conf. Series: Materials Science and Engineering 1039 (2021) 012014 doi:10.1088/1757-899X/1039/1/012014

# Study of calcium-silicate composite lightened by waste expanded perlite

**P Sebestova**[1], **V Cerny**[1] **and R Drochytka**[1]

[1] Brno University of Technology, Faculty of Civil Engineering, Veveri 331/95,602 00 Brno, Czech Republic

Email: sebestova.p@fce.vutbr.cz

**Abstract**. The term calcium-silicate composite can be recalled several basic building materials. The most common representatives are lime-silicate bricks and autoclaved aerated concrete. It is any material based on the reaction of lime and siliceous raw material. In this paper, attention is paid to composites based on autoclaved aerated concrete production technology. Possibilities of lightening the composite utilizing waste expanded perlite as a secondary raw material were investigated without the use of aluminium powder, which is commonly used in aerated concrete technology. Replacements of 10%, 30%, 50%, and 100% waste expanded perlite for sand were used. Based on physico-mechanical properties, and with the assessment of the microstructure, it has been found that up to 50% of the replacement of waste expanded perlite results in a lightening of the composite while increasing the compressive strength.

## 1. Introduction

The conventional substitution of quartz sand in autoclaved aerated concrete, whether partial or complete, if it would be possible to reduce the bulk density by substitution of sand with waste expanded perlite (WEP) and lightening of the composite without the use of aluminum powder.

As mentioned above, the calcium-silicate composite of interest is based on the autoclaved aerated concrete production technology. For clarity, this technology will be briefly described. The basic raw materials to produce autoclaved aerated concrete are lime, silica sand, and water. Cement, gypsum, and aluminium powder are also used. Lastly, other additives may be used, depending on the manufacturer. The raw materials are mixed to form a very fluid mixture to which aluminium powder is added at the end of the mixing. Then, the liquid mixture is poured into the mold. During the reaction of the aluminium powder with lime and water, hydrogen gas is produced, thereby aerating the liquid mixture. In this way, the porous concrete obtains its typical porous structure. After demolding and cutting the green body mass to the desired size, the intermediate is ready for curing. Curing performed in an autoclave at 180–190°C for 7–12 hours and at a pressure of 1.2–1.4 MPa (the temperature, time, and pressure of autoclaving vary by manufacturer). During curing, chemical reactions take place in the microstructure of the autoclaved aerated concrete, giving the resulting products their strength. After curing, the products are immediately ready for dispatch [1–3].

To understand the changes in microstructure, it is first necessary to describe the basic mechanisms of calcium hydro silicate (CSH) formation. The basic raw materials for the production of calcium-silicate composites are lime and quartz sand. These materials are characterized by oxide, they are primarily composed. Lime is a carrier of calcium oxide (CaO), silica sand contains a high amount of silica dioxides ($SiO_2$). The last important component for the formation of the CSH phase is water. However, the formation and interconnection of these three components would not have been so simple after mixing. The silica dioxides contained in the sand are in crystalline form, i.e. it is insoluble and under normal conditions, its reaction is very weak, and only on the grain surface. Therefore, it is necessary to cure the composites under hydrothermal conditions. During curing, $SiO_2$ reacts with CaO and $H_2O$ to form a CSH phase. However, the CSH phases are many forms, and each form depends on the specific temperature, cure time, and ratio of individual oxides. For calcium-silicate composites, the CSH phase is an important 1.1 mm tobermorite. It is a layered mineral in the form of long needles, which connects individual grains of unreacted sand and ensures the strength of the product [4–7].

The effect of WEP on the calcium-silicate composite is a subject almost unexplored. Perlite is an amorphous volcanic rock that can be classified as a natural pozzolanic material due to its glassy structure, high silica dioxides, and alumina oxide content [8]. It is these two oxides, alumina oxide and silica dioxide, which are important for the use of WEP in the technology of autoclaved calcium-silicate composites. During perlite formation by the rapid cooling of ejected material in the water medium, about 5% of the water was bound to the rock, which is used in the production of expanded perlite. Perlite is heated to a temperature in the range of 900–1000°C, which converts the contained water into steam. As a result of the expansion of the material, expanded perlite is formed which has up to 4–20 times higher volume [8–11]. This also increases the specific surface area, so expanded perlite is a very porous material and thus becomes a good thermal insulator. WEP is characterized by low bulk density, fire resistance, and resistance to fungi and algae [9]. During the production of expanded perlite, WEP is formed, which belongs to the waste category "O" [12][13]. It can be further recycled, used as a secondary raw material, or deposit in construction waste dumps. Kotvica et al. [13] utilized ground expanded perlite as an additive to cement and increased compressive strength by up to 50% at 35% substitution. Further research has found that WEP is suitable for zeolite synthesis and suppresses the alkaline silica reaction when used in cement [13, 14]. In autoclaved aerated concrete, WEP was used as a substitute for quartz sand, and a 10% substitution resulted in a 15% reduction in thermal conductivity [1]. The possibility of using WEP is also confirmed by other studies dealing with the use of expanded perlite in autoclaved aerated concrete and fly ash-lime-gypsum mixture [15][16].

This experiment aimed to determine and define possibilities of lightening of autoclaved calcium-silicate composite with light filling. Waste expanded perlite (WEP) was chosen as a light filler. The data should then serve for further research into the use of lightweight filler in both calcium-silicate composite and autoclaved aerated concrete.

## 2. Materials and methods

### *2.1. Materials*

Quartz sand, quicklime, cement, gypsum, and water were used to prepare the reference sample. Quartz sand (CZ, Dolní Lutyně) contains more than 92% silica dioxides. It can, therefore, be described as a very pure material. Quicklime (CZ, Beroun) is labeled CL 90-Q (R5, P1) according to ČSN EN 459-1 "Building quicklime". Portland cement has the designation CEM I 52.5 N (SK, Rohožník). Flue gas desulfurization (FGD) gypsum contains at least 95% calcium sulfate (CZ, Dětmarovice).

**Table 1.** Raw materials mixture.

| Raw materials | REF [%] |
|---|---|
| Lime | 9 |
| Cement | 14 |
| FDG Gypsum | 3 |
| Quartz | 75 |

**Table 2.** Chemical composition of raw materials.

| Raw materials | Oxide | CaO | MgO | $K_2O$ | $Al_2O_3$ | $P_2O$ | $Fe_2O_3$ | MnO | $Na_2O$ | $SiO_2$ | $TiO_2$ |
|---|---|---|---|---|---|---|---|---|---|---|---|
| WEP | [%] | 1.26 | 0.23 | 4.56 | 13.10 | 0.02 | 2.02 | 0.06 | 2.14 | 74.00 | 0.16 |
| Sand | [%] | 0.23 | 0.16 | 1.53 | 2.53 | 0.01 | 0.84 | 0.01 | 0.70 | 92.91 | 0.14 |
| Cement | [%] | 64.25 | 2.02 | 3.32 | 5.40 | 0.16 | 3.06 | 0.10 | 0.36 | 19.67 | 0.31 |
| Lime | [%] | 95.62 | 0.79 | - | - | - | - | - | - | - | - |

The raw materials mixtures are based on the autoclaved aerated concrete technologies but without the use of aluminium powder (table 1). The chemical composition of raw materials is given in table 2.

Quartz sand was substituted by 10%, 30%, 50%, and 100% WEP in raw materials mixtures. The substitution was not in weight units, as is usual with other raw materials. Due to the very low bulk density of WEP, which was set at 300 kg/m$^3$, it was necessary to choose a substitution in volume units. The determined water-absorbing capacity of WEP was 260%. The chemical composition of WEP (CZ, Šenov u Nového Jičína) is given in table 2. The phase composition of WEP is quartz, cristobalite, muscovite, biotite, anorthite sodian, albite, and anorthoclase. The particle size distribution of WEP and quartz sand is given in Figure 1. Quartz sand contains the sand contains a higher proportion of fine grains. The curves have a similar increase. For this reason, the grain size distribution will be similar for WEP and quartz sand.

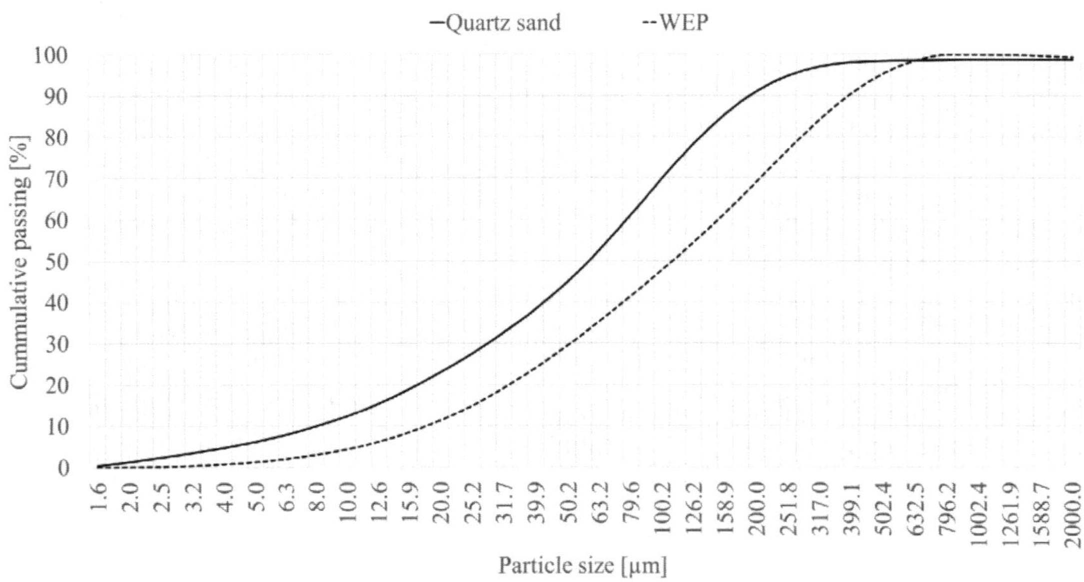


**Figure 1.** The grain size of WEP and quartz sand.

## *2.2. Preparation of mixture*

The sand and gypsum were homogenized in a mixer separately. After careful mixing, water was added. Reactive components (lime, cement) and WEP were mixed into this mixture. the prepared liquid mixture was poured into molds of 40 × 40 × 160 mm. The flow test was immediately determined on the remaining mass. WEP is a very absorbent material. With such a high-water absorption, WEP can take up a large amount of water during mixing, which affects the resulting consistency of the mixture. To monitor the effect of the amount of WEP in the composition on its consistency, the water was constant.

After 24 hours maturing of the mixture in the molds, the samples were removed from the molds and placed in a laboratory autoclave. Curing was carried out at 190°C for 7 hours.

*2.3. Methods*
Basic physical-mechanical properties, such as compressive strength and bulk density according to ČSN EN 1351 and ČSN EN 678, were determined on the samples.

The microstructure of samples was studied by X-ray diffraction analysis and scanning electron microscope. The method of X-ray diffraction analysis is based on the fact that the crystal lattice of each mineral has a specific distance between two parallel surfaces. This value can be obtained at a known wavelength $\lambda$ and a measured reflection angle $\theta$. To be able to subject X-ray diffraction analysis, it was necessary to grind to a size of less than 20 μm in a so-called micromills. X-ray powder data were collected on an Empyrean PANalytical using Cu-cathode, $\lambda = 1.540598$ for K$\alpha$ 1. Angular reproducibility is <0,0002°. For scanning by the electron microscope, it was again necessary to obtain central cores of samples, without contamination by surrounding impurities. These samples were then gold plated to ensure their conductivity. They were then inserted into the microscope body to remove residual moisture and air. Samples were analyzed on a TESCAN MIRA 3 XMU scanning electron microscope at an accelerating voltage of 20 kV.

The flow test is performed using a cylinder (height 5.4 cm, diameter 7.1 cm) and a plastic plate. The mixture is poured into the cylinder. The cylinder is raised and immediately after spilling the mixture, two perpendicular diameters are measured. The results are given in cm.

## 3. Results

*3.1. Physical-mechanical properties*
Figure 2 shows the physico-mechanical properties of samples with WEP substitution. The graph shows the dependence of compressive strength and flow of the mixture on bulk density. Each curve shows one dependence, namely the compressive strength on the bulk density and the flow of the mixture on the bulk density. The decrease in bulk density in the graph is dependent, inter alia, on the increasing amount of WEP substitution. The increasing WEP substitution value moves from right to left in the graph. The table with the values used is also shown in the graph.

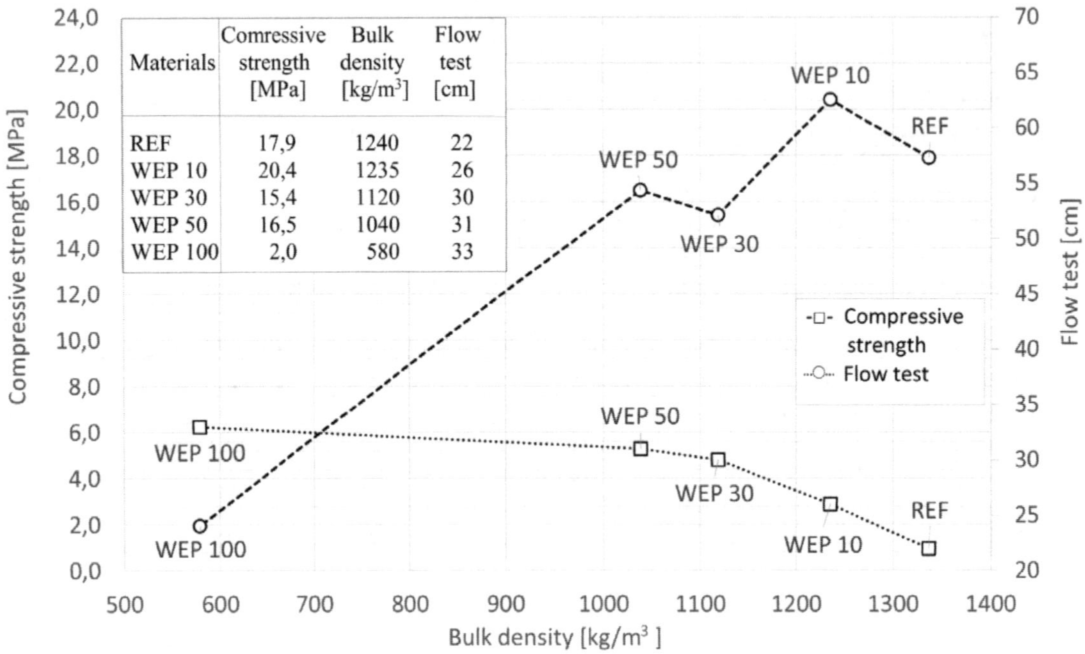


| Materials | Comressive strength [MPa] | Bulk density [kg/m$^3$] | Flow test [cm] |
|---|---|---|---|
| REF | 17,9 | 1240 | 22 |
| WEP 10 | 20,4 | 1235 | 26 |
| WEP 30 | 15,4 | 1120 | 30 |
| WEP 50 | 16,5 | 1040 | 31 |
| WEP 100 | 2,0 | 580 | 33 |

**Figure 2.** Physical-mechanical properties of calcium-silicate composite with substitution of WEP.

Interesting results were obtained when measuring the flow of the mixture. WEP has a high-water absorption capacity (260%) and it was expected that with higher sand substitution of WEP the flow of the mixture would decrease. However, the liquid mixture became more fluid. To elucidate the increase in the fluidity of the mixture, optical images if WEP were determined. In the images of WEP in water, an air layer formed around the WEP grain, which prevents water from penetrating its structure. The air layer disappears within a few minutes, creating only the apparent fluidity of the mixture. Therefore, a longer and more intensive mixing of the mixture is suggested.

One of the basic parameters of lightweight calcium-silicate composites is the bulk density. The calcium-silicate composite is characterized by a bulk density of 1450 kg/m$^3$. When WEP is used, the material is gradually lightened. Significant weight loss was recorded at 100% sand substitution by WEP.

The unexpected result was the achieved compressive strength of lightweight calcium-silicate composites. Up to 50% sand substitution, composites were lightened, and strength increased. At 100% sand substitution by WEP, the strength decreased to 0.6 MPa. However, considering the significant decrease in bulk density, the resulting calcium-silicate composite can be described as a potential thermal insulator.

### *3.2. Microstructure*

As described in the introduction, it is necessary to identify newly formed minerals to explain the behavior of the properties of calcium-silicate composites. In this case, the focus is on the mineral 1.1 mm tobermorite, which is a carrier of strength. Images of the minerals are shown in Figure 4 and figure 4.

**Figure 3.** SEM images of the reference sample.

The calcium-silicate composite contains plate-like 1.1 mm tobermorite crystals. This shape of tobermorite is most often found in the crystalline quartz reaction. Tobermorite of composite with 10% sand substitution by WEP does not change its shape. In addition, a non-crystalline CSH gel is present in the sample. At 30% and 50% sand substitution by WEP, the shape „house of cards" of 1.1 nm tobermorite described by Chucholowsky [17] occurs in his work. This shape of tobermorite is characterized by the support of compressive strengths, which has been manifested here. The composite with 100% substitution by WEP does not contain 1.1 nm tobermorite crystals. It contains long hairy new phases. These phases are called α-$C_2SH$ and are formed as the first reaction product using 0.10–1 mm grain silicon sources [18].

**Figure 4.** SEM images of calcium-silicate composite with sand substitution by WEP.

The microstructure of the composite can also be determined by X-ray diffraction analysis. In this analysis, it is possible to determine the degree of crystallization of the new phases using the intensity of the diffraction line. This analysis was also used in the research of expanded calcium-silicate composite. The hydrothermal conditions were adjusted such that the quartz surface was partially dissolved and released $H_3SiO_4^-$ or $H_2SiO_4^{2-}$ ions into the alkaline solution. These ions react with $Ca^{2+}$ ions present in the alkaline solution to form crystalline calcium hydrogen silicate phases. For this reason, crystalline 1.1 mm tobermorite or other crystalline phases should appear in all samples. This assumption partly correlates with the results presented in Figure 5.

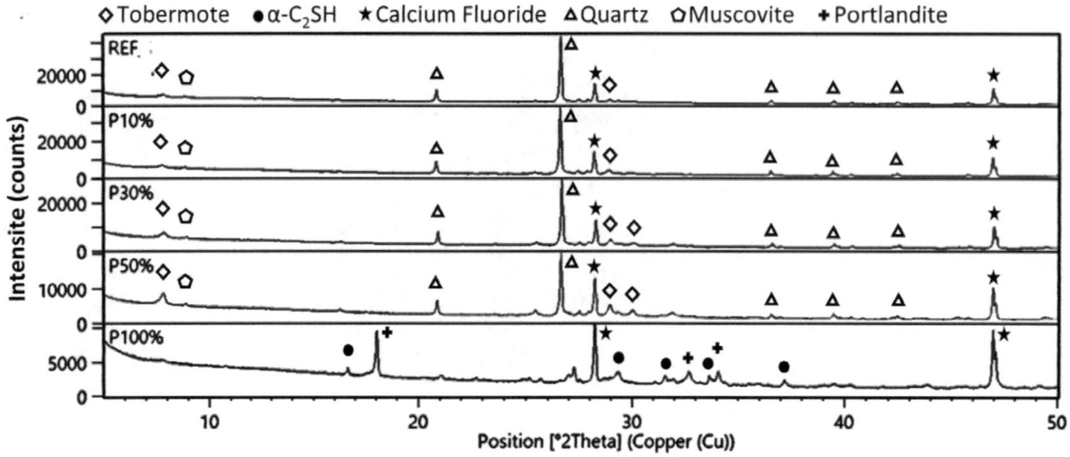


**Figure 5.** X-ray diffractograms of calcium-silicate composite with sand substitution by WEP.

Samples with the 10–50% sand substitution by WEP contain 1.1 mm tobermorite. In these samples, WEP increased the crystallization rate and the crystallization of tobermorite was most intense at 50% substitution. These samples also contain quartz and muscovite. These minerals are contained in the sand. Calcium Fluoride was added to the samples as standard.

1.1 mm tobermorite was not formed in the sample with 100% sand substitution by WEP. Quartz and muscovite are not present in the sample. Quartz and muscovite contain sand. Therefore, these minerals are not present in the sample with 100% sand substitution by WEP. However, there are other new phases in the sample. The sample contains $\alpha$-$C_2SH$ and portlandite. The results confirm the results of the SEM analysis.

## 4. Discussion

### *4.1. Structure formation used crystalline silica dioxides (sand)*

In the case of the reference sample, which was formed using sand, that is, crystalline silica dioxides, it can be stated that this was a process of microstructure formation already described. As a result, a plate-like 1.1 mm tobermorite was formed which crystallized on the surface of the sand grains to form a solid bonding matrix between these grains. This is related to the achieved compressive strength, which is standard for a calcium-silicate composite.

### *4.2. Structure formation used crystalline silica dioxides (sand) and amorphous silica dioxides (WEP)*

The combination of crystalline and amorphous silica dioxides, the formation of 1.1 mm tobermorite is inclined to the side of the higher amount of amorphous silica oxides. In previous studies [19][20], the positive effect of amorphous silica dioxides on the formation of 1.1 mm tobermorite was confirmed. The results obtained to confirm this statement. Another reason for the increased crystallization of tobermorite is in the porous character of WEP. The respective surface and hence the reaction surface of silicon sources makes a decisive role in the solubility behavior of silicon. Due to the high reaction surface, WEP dissolves to a greater extent and thus contributes to the crystallization of 1.1 mm tobermorite.

### *4.3. Structure formation used amorphous silica dioxides (WEP)*

In the case of a sample with 100% sand substitution by WEP, unlike other samples, a crystalline 1.1 mm tobermorite was not formed. SEM and XRD analyses confirmed the presence of $\alpha$-$C_2SH$, which is the primary curing product and is partially crystalline [18]. During later curing, $\alpha$-$C_2SH$ decomposes in favor of 1.1 mm tobermorite. However, in this sample, only the $\alpha$-$C_2SH$, portlandite and CSH gel were formed which is evident in Figure 5. There may be several reasons why tobermorite is not formed.

The high curing temperature and reactivity of the silica dioxide cause the formation of a supersaturated solution. Increased dissolution changes the entry conditions for tobermorite formation. [21], [22] This phenomenon is also confirmed by the presence of portlandite. Portlandit is not involved in the reaction in its entirety, like other samples. Another explanation is in exceeding the curing temperature (180°C). The solubility curve according to states that the solubility curves of portlandite and amorphous silicon dioxides intersect at 120°C [22]. At higher temperatures, the solubility of portlandite decreases while the solubility of amorphous silica dioxides increases. As a result of this rule, the solution could be supersaturated in favor of silica dioxides. WEP is also characterized by porosity and thus generates a large reaction area which, in turn, may have resulted in a change in the entry ratio of calcium oxide to silica dioxides. According to studies, the ideal ratio for 1.1 mm tobermorite formation is 0.83, and an unstable calcium hydrogen silicate phase is formed at a higher ratio.

## 5. Conclusion

Nowadays, it is the duty to address sustainable technologies, environmental solutions, and efforts to recycle and use as much waste as possible. This study focused precisely on the use of waste material that waiting for its use. WEP is tested in the production of calcium-silicate composite. Research has shown that the recovery of this waste can have a positive effect on the properties of the composite. By using WEP, the compressive strength of composites was increased, and the bulk density of composites was reduced, too. At 100% sand substitution by WEP, a composite with a very low bulk density of 640 kg/m$^3$ and a compressive strength of 0.6 MPa was formed.

**Acknowledgment**

This paper was elaborated within project No. [FV30327] " Progressive high-quality autoclaved aerated concrete non-waste technology with usage of renewable resources", and was also funded under [FAST-J-20-6165] "Research of substitution of binder component of lime-silicate composites using alternative raw materials".

**References**

[1] Ling T, Mo K H, Qu L, Yang J and Guo L 2017 Mechanical strength and durability performance of autoclaved lime-saline soil brick *Const. Buil. Mat* **146** pp 403–409

[2] Dachowski R and Nowek M 2016 Landfill leachate as an additive in sand-lime products *Proc. Eng.* **161** p 572

[3] Pytel Z and Malolepszy J 2000 Effect of mineral admixtures on some properties of sand-lime bricks *Elsevier* pp 371–382

[4] Kurama H, Topcu İ B and Karakurt C 2009 Properties of the autoclaved aerated concrete produced from coal bottom ash *Jour. Mat. Proc. Tech.* **209** pp 767–773

[5] Zou J, Guo CH, Jiang Y, Wie C and Li F 2016 Structure, morphology and mechanism research on synthesizing xonotlite fiber from acid-extracting residues of coal fly ash and carbide slag *Mat. Chem. Phys.* **172** pp 121–128

[6] Mitsuda T 1982 Influence of starting materials on the hydrothermal reaction in the $CaO$-$SiO_2$-$H_2O$ system *Jour. Jap. Assoc. Mineral.* **77** pp 317–329

[7] Rozycka A and Pichor W 2016 Effect of perlite waste addition on the properties of autoclaved aerated concrete *Constr. Buil. Mat.* **120** pp 65–71

[8] Kotwica Ł, Pichor W and Nocun-Wczezlik W 2016 Study of pozzolanic action of ground waste expanded perlite by means of thermal methods *Jour. Ther. Anal. Calor.* **123** pp 607–613

[9] Expandovaný perlit *Perlit.cz* (*Online* http://www.perlit.cz/expandovany-perlit/)

[10] Long W J, Tan X W, Xiao B X, Han N X and Xing F 2019 Effective use of ground waste expanded perlite as green supplementary cementitious material in eco-friendly alkali activated slag composites *Jour. Cl. Prod.* **213** pp 406–414

[11] Sengul O, Azizi S, Karaosmanoglu F and Tasdemir M A 2011 Effect of expanded perlite on the mechanical properties and thermal conductivity of lightweight concrete *Ener. Buil*

**43** pp 671–676
[12] Expandovaný perlit: Bezpečnostní list *Perlit.cz* (*Online* http://www.perlit.cz/wp-content/uploads/2017/03/bezpecnosti-llist_EP.pdf)
[13] Kotwica Ł, PICHÓR W, Kapeluzsna E and Rozycka A 2017 Utilization of waste expanded perlite as new effective supplementary cementitious material *Jour. Cl. Prod.* **140** pp 1344–1352
[14] Singh M and Garg M 1991 Perlite-based building materials - a review of current applications *Const. Buil. Mat.* **5** pp 75–81
[15] Ayudhya B I N 2011 Compressive and splitting tensile strength of autoclaved aerated concrete containing perlite aggregate and polypropylene fiber subjected to high temperatures *Son. J. Sci. Technol.* **33** pp 555–563
[16] Demir I and Serhat Baspinar M 2008 Effect of silica fume and expanded perlite addition on the technical properties of the fly ash-lime-gypsum mixture Const. Buil. Mat. **22** pp 1299–1304
[17] Chucholowski C, Holger H and Thienel K 2018 Improving the recyclability, environmental compatibility, and $CO_2$ balance of autoclaved aerated concrete by replacing sulfate carrier and cement with calcined clays *Ce/papers* **2** pp 503–512
[18] Harmann A 2004 Untersuchungen zum Kristallisationsverhalten und zur Morphologie von 11 Å Tobermorit in Abhängigkeit von der Reaktivität der Kieselsäurequelle und dem Ionenbestand der Hydrothermallösung *Fach. Geowiss. Geog. Univer.* (Hanover)
[19] Ma B G, Cai L X, Jian S W and Su L 2014 Influence of Fly Ash Fineness on Autoclaved Energy Conservation Block's Performance and Hydration Products *Key Eng. Mat.* **599** pp 302–309
[20] Serhat Baspinar M, Demir I, Kahraman E and Gorhan G 2014 Utilization potential of fly ash together with silica fume in autoclaved aerated concrete production *KSCE Jour. Civil Eng.* **18** pp 47–52
[21] Dilmore R M and Nneufeld R D 2001 Autoclaved aerated concrete produced with low $NO_x$ burner/selective catalytic reduction fly ash *Jour. Ener. Eng.* **127** pp 37–50
[22] Isu N, Teramura S, Ishida H and Mitsuda T 1995 Influence of quartz particle size on the chemical and mechanical properties of autoclaved aerated concrete (II) fracture toughness, strength and micropore *Cem. Noc. Res.* **25** pp 249–254
[23] Hartmann A, Pawlas G, Petrow V and Buhl J C 2014 Dissolution kinetics and phase transformations of autoclaved aerated concrete in alkaline and acid media without further additives *Jour. Sol. Waste Tech. Man.* **40** pp 97–109

# Utilization of flue dust arising from metakaolin production for alkali-activated materials preparation

**D Kubatova[1], B Nespor[1], I Kukletova[1] and R Cechmanek[1]**

[1] Research Institute for Building Materials, Hnevkovskeho 65, 617 00, Brno, Czech Republic

Email: kubatova@vustah.cz

**Abstract**. This paper assesses the feasibility of using flue dust, which is a residue from metakaolin manufacturing in a rotary kiln as a raw material for production of alkali-activated materials. This material contains besides metakaolin also residues of clays, and mullite. Metakaolin and blast furnace slag were used as initial materials. Flue dust was used for the replacement of initial materials up to 100%. Mortars were prepared for the determination of compressive strength by adding sand at the weight ratio of 1:1. As an alkaline activator was used potassium water glass and s/w ratio was 1.25. The prepared samples were tested for physical-mechanical properties after 28, 56, 90 and 180 days. Setting time and leaching characteristic were also evaluated.

## 1. Introduction

Geopolymer binders are based on aluminosilicate materials and possess good mechanical properties. Thanks to these properties, geopolymers can be considered as an alternative binder compared to Portland cement. In earlier studies, metakaolin was used as the main aluminosilicate raw materials, due its high reactivity and the formation of homogenous reaction product [1]. Most of the waste materials such as fly ash, blast furnace slag and various mining waste can be used for the alkaline reaction. The type and nature of initial materials used directly affect the final properties of geopolymer prepared from these materials [2]. Many industrial wastes have been tested to reduce the cost of geopolymers. The type and nature of the initial materials used directly affect the final physical and chemical properties of geopolymer prepared from waste materials.

In the previous work [3] there were studied properties of geopolymers synthesized from the clay calcined in a rotary kiln. The clay contained 30–70% of kaolinite which was transformed by calcination at 750°C to metakaolinite.

The thermally treated kaolin mining wastes formed mainly kaolinite were tested [4]. Kaolinite-based clays require a thermal activation and therefore it raises cost of these materials [5]. It has been proved that calcined materials which are mostly amorphous, usually display a higher reactivity during geopolymerization compared to non-calcined materials [6].

The use of raw kaolin itself in geopolymer has not been studied in detail. Heah et al [7] investigated kaolin-based geopolymers in sodium alkaline-activator and confirmed good volume stability of these materials. Various combinations of kaolinite, albite and fly ash have been tested for the preparation of geopolymers. The results showed that the compressive strength was greatly increased when these three minerals undergo geopolymerization together [8].

The aim of this work was to prove the possibility of using flue dust in its original state without sufficient thermal activation. This material has already undergone a calcination process in a rotary kiln, and it contains besides metakaolin also residues of kaolinite. Physical observation, workability and leachability test of alkali-activated materials are reported in this paper.

## 2. Experimental part

### *2.1. Materials*

Metakaolin (MK) and blast furnace slag (BFS) were used as main initial materials. MK Mefisto L05 was supplied by České lupkové závody, a.s. (Czech Republic), finely ground BFS originated from Třinec (Czech Republic). Rotary kiln flue dust (RON) is a by-product of metakaolin production. Chemical composition of all raw materials was determined by XRF analysis (PANalytical Axios PW4400/30) (See table 1). Characteristic particle diameters $D_{10}$, $D_{50}$ and $D_{90}$ are summarized in table 2. Potassium water glass was used as alkaline-activator (Ms = 1.78).

**Table 1.** Chemical composition of materials used in mixtures (%).

| | $SiO_2$ | $Al_2O_3$ | CaO | $Na_2O$ | $K_2O$ | MgO | $Fe_2O_3$ | $TiO_2$ | $SO_3$ | LOI |
|---|---|---|---|---|---|---|---|---|---|---|
| MK | 53.21 | 42.94 | 0.20 | 0 | 0.78 | 0.28 | 0.92 | 1.56 | 0.05 | 2.10 |
| BFS | 39.99 | 8.99 | 38.70 | 0.47 | 0.48 | 8.71 | 0.49 | 0.34 | 1.84 | 0.71 |
| RON | 52.06 | 39.98 | 1.16 | 0.28 | 0.88 | 0.71 | 2.19 | 2.04 | 0.58 | 5.55 |

**Table 2.** Particle diameters of used materials (μm).

| | $D_{10}$ | $D_{50}$ | $D_{90}$ | Mean diameter |
|---|---|---|---|---|
| MK | 1.21 | 4.65 | 11.00 | 5.52 |
| BFS | 1.66 | 10.73 | 28.63 | 13.38 |
| RON | 2.06 | 16.31 | 58.23 | 24.08 |

The XRD analysis of these raw materials was performed. The amorphous phase was a main component of MK. MK contained 83% of amorphous phase, quartz ($SiO_2$), kaolinite ($Al_2Si_2O_5(OH)_4$), illite ($KAl_2(AlSi_3O_{10})(OH)_2$), anatase ($TiO_2$) and mullite ($Al_6Si_2O_{13}$). BFS contained amorphous phase (78%), calcite ($CaCO_3$), than merwinite ($Ca_3MgSi_2O_8$), akermanite ($Ca_2MgSi_2O_7$) and larnite ($Ca_2SiO_4$). RON contained 64% of amorphous phase, 17% of kaolinite, also quartz ($SiO_2$), anatase ($TiO_2$), illite ($KAl_2(AlSi_3O_{10})(OH)_2$) and mullite ($Al_6Si_2O_{13}$).

The composition of raw materials and final mortars was determined by X-ray powder diffraction. The X-ray diffraction analysis was conducted using Bruker D8 Advance apparatus with Cu anode ($\lambda K\alpha = 1.54184$ Å) and variable divergence slits at Θ-Θ Bragg-Brentano reflective geometry.

### *2.2. Sample preparation*

Metakaolin and blast furnace slag were used as the primary aluminosilicate raw material. Ratio $SiO_2/Na_2O$ was 6.9 and ratio $SiO_2/Al_2O_3$ was 4.28. MK and BFS was partially replaced by RON at 0–100 wt. %. After homogenization, the solid materials were mixed with alkaline-activator in ratio S/L 1.25. Shale (0-0.4 mm) was used as a filler in a weight ratio of 1:1. The mortars were casted into moulds sized 40 × 40 × 160 mm. The samples were then wrapped with a plastic foil to avoid moisture evaporation. After 24 h, hardened specimens were demoulded and stored in plastic bags under laboratory conditions for 180 days.

### 2.3. *Testing methods*

The consistency of prepared mortars was determined according to EN 12350-5 using the flow table test, where the mean diameter of a test sample was measured. Initial setting time of geopolymer pastes was determined using the Vicat needle method according to EN 196-3:2016.

The flexural strength of a hardened mortar was determined by three point loading of a prism specimen, subsequently to the failure and breakage of this specimen the compressive strength was determined on each half of the prism 160 × 40 × 40 mm. The samples were tested in several terms up to 180 days.

The leachability test was performed after 28 days at 20°C. The samples were soaked in water up to 7 days. The volume of water was determined by calculated surface of the tested prism at the volume ratio of 1:5. Once a day the leachate was stirred. The resulting leachate was analysed by ICP-OES method (ICP IRIS Intrepid II XSP Duo) and $K^+$, $Na^+$, $Ca^{2+}$ and $SO_4^{2-}$, $NO_3^-$, $Cl^-$ were determined.

## 3. Results and discussion

### 3.1. *Workability and setting time*

In the first phase, the effect of the amount of RON in the mixture on the consistency and initial setting time was studied. Figure 1a and figure 1b show the consistency and setting times of the pastes as a function of the RON content.

A small amount of the raw material RON results in an increase in the mortar diameter, then decreases up to 140 mm for the 100RON sample. It is clearly illustrated that the setting times were prolonged of various degrees by adding RON. The setting time was slowly affected by RON up to 40%. At higher dosages, the onset of setting time significantly prolonged. The reference sample (0RON) has a setting time of about 115 min, while 100RON over 2,100 min.

One of the factors that fundamentally affect the workability in particle size of starting materials. RON has a higher ratio of coarser particles than metakaolin and slag (See table 2). Setting time of geopolymer mixtures depends on the specific surface area. The lower specific surface area leads to higher setting time of geopolymer mixtures [9, 10]. Simultaneously, the kaolin particles should decrease workability due to the plate-like structure [7]. The kaolinite content in RON results in lower reactivity of mixtures due to kaolinite slower dissolution in alkaline solution [11].

a)

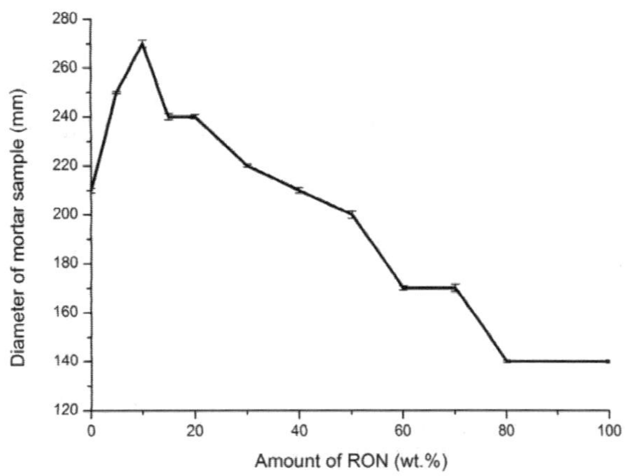

b)

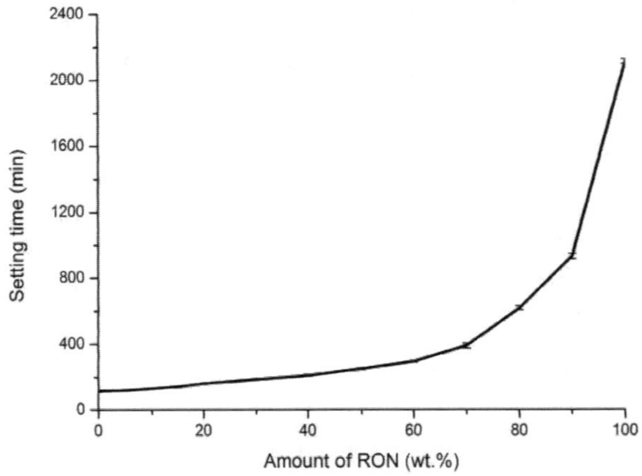


**Figure 1.** Results of consistency determination (a) and initial setting time (b) depending on the RON content in the mixtures.

### *3.2. Compressive and flexural strength*

The compressive strength and flexural strength of geopolymer mortars at the ages of 28, 56, 90 and 180 days are depicted in figure 2. The strengths decreased as the amount of RON in the mixture increased. The compressive strength of the reference sample (0RON) was 54 MPa and flexural strength was 7 MPa after 28 days. The 100RON sample had much lower compressive and flexural strength after 28 days, i.e. 20 MPa and 3.8 MPa, respectively. Overall results showed that strengths increased with the ageing time. The most significant increase in strength was recorded within 90 days. After 180 days the strength increases slightly especially in samples containing RON. After 180 days the reference sample had 72 MPa and 8.8 MPa, respectively, and 100RON sample had 41.6 MPa and 6.5 MPa, respectively.

Strengths were strongly affected by the composition of samples. The kaolinite from RON required more time for reaction and became hard solid. Because of the low reactivity of kaolinite, a weak structure was formed resulting in the decrease of strength [7, 8]. In addition to composition of the raw materials, strengths can also be affected by particle size. Coarser particles could harder fill the spaces leading to less dense structure [12].

Heah [7] was involved in the preparation of kaolin-based geopolymer with sodium alkaline-activator with various solid-to-liquid ratios. Compressive strengths exceeding 5 MPa were measured for the prepared samples. In another work [8], geopolymers were prepared from fly ash, which was partially replaced by kaolinite. In this case, a compressive strength of about 30 MPa was achieved, using a sodium alkaline activator. Using kaolinite as the source of aluminium definitely had a negative effect on the compressive strength as compared to that of the matrices where metakaolinite was used.

a)

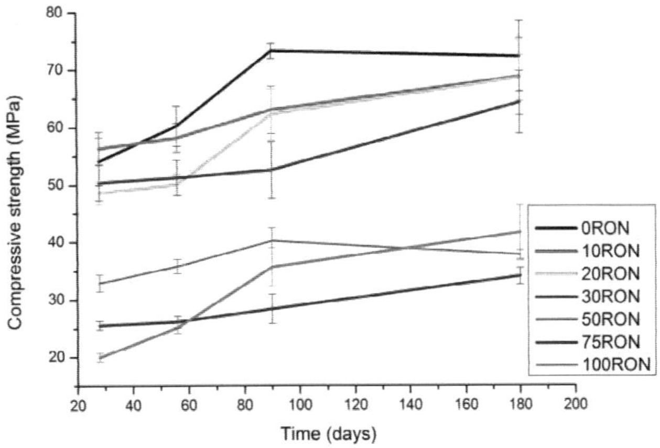


b)

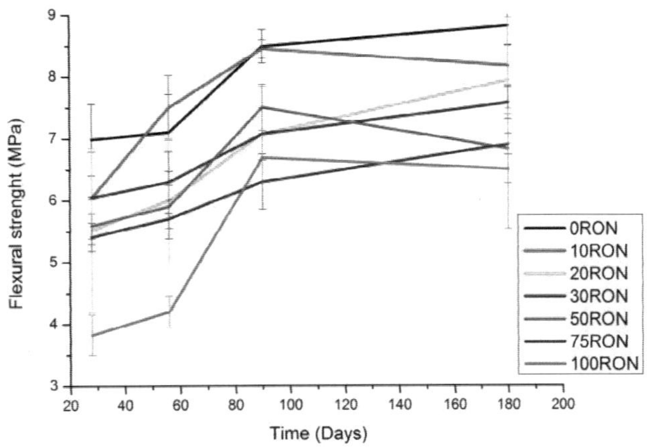


**Figure 2.** Development of compressive strength (a) and flexural strength (b) in time.

### *3.3. XRD measurement*

The structure analysis of prepared pastes after 28, 90 and 180 days was carried out by XRD method. The XRD patterns of all samples had a broad hump, characteristic for an amorphous phase. In the case of reference sample without RON addition, the main crystalline phases were calcite ($CaCO_3$), quartz ($SiO_2$), anatase ($TiO_2$), muscovite ($KAl_3Si_3O_{12}H_2$), mullite ($Al_6Si_2O_{13}$), merwinite ($Ca_3MgSi_2O_8$) and akermanite ($Ca_2MgSi_2O_7$). Kaolinite characteristic peaks (7.19 Å; 1.48 Å; 3.58 Å) were identified in all samples containing RON (See figure 3a). Kaolinite is unable to be incorporated into the polymer structure due to fact, that the dissolution of kaolinite in alkaline solution is extremely slow [6, 7]. The intensity of kaolinite peak increased with the amount of RON in the matrix and decreased with ageing days. After 180 days, kaolinite peaks were still identified in the XRD pattern of all samples where the raw materials have been replaced with RON. Calcite-related peaks (3.03 Å) from slag were not identified in the 100RON sample (See figure 3b).

RON content in the matrix influenced the formation of precipitate on the surface of the samples. This was especially evident in the samples 75RON and 100RON after 90 and 180 days. These precipitates were identified as aragonite ($CaCO_3$) by X-ray diffraction.

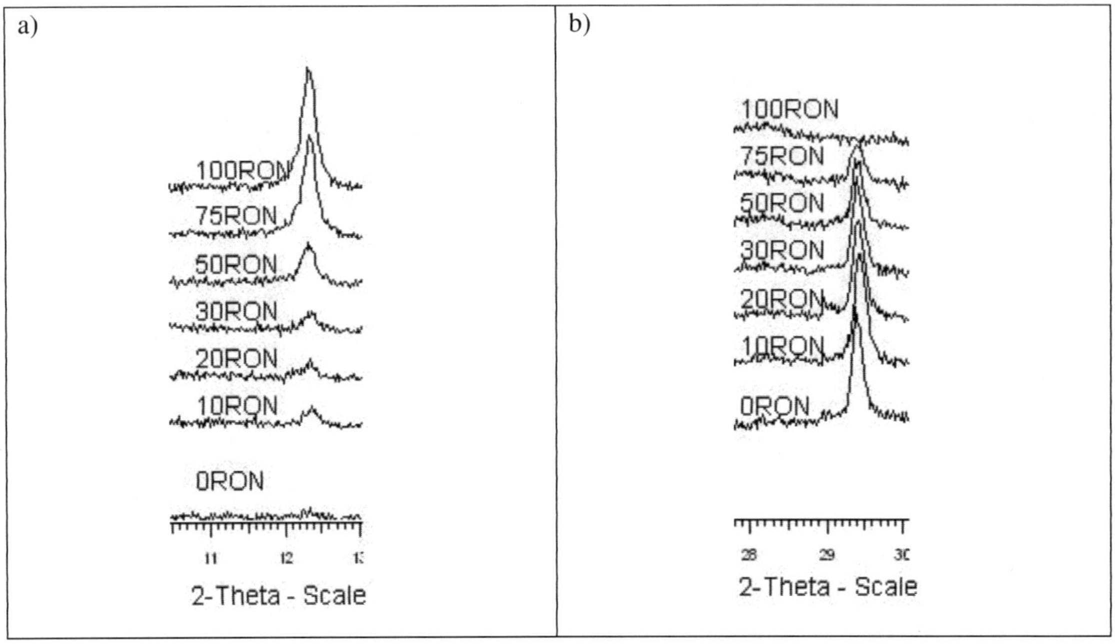


**Figure 3.** XRD pattern of samples after 180 days – a) peak of kaolinite (7.14 Å) b) peak of calcite (3.03 Å).

### *3.4. Leaching test*

After 28 days of maturation, leaching tests were carried out and ions that could potentially contribute to efflorescence were analyzed. Using ICP-OES method $K^+$, $Na^+$, $Ca^{2+}$ and $SO_4^{2-}$, $NO_3^-$, $Cl^-$ were determined. The use of potassium water glass should reduce the possibility of efflorescence. The development of determined ions in leachate is shown in figure 4a and 4b.

The content of chlorides and nitrates in the extract was very low as well as the content of sodium and calcium ions. Conversely, the content of sulfate and potassium ions in the leachate increased very intensively with the amount of RON in the matrix. The results suggest that potassium ions are released into the solution because there is no appropriate structure of the geopolymer. The work by Jaarsveld [2] confirmed that the inclusion of kaolinite serves to provide structure-forming species to the overall geopolymerization process, but a large amount of added kaolinite may not take part in the synthesis reaction at all. The samples 75RON, 100RON were disintegrated after 7 days of soaking. This is probably due to the very weak structure as shown by the very low compressive strength.

An increasing concentration of sulfate ions could probably cause aragonite to precipitate on the sample surface. Previous work [13] has confirmed that a pH below 11 is suitable for crystallization of aragonite. Aragonite can be formed from sodium carbonate and calcium sulfate. Aragonite formation is promoted by the addition of sodium sulfate due to a decrease in supersaturation and sulfate ion adsorption.

a)

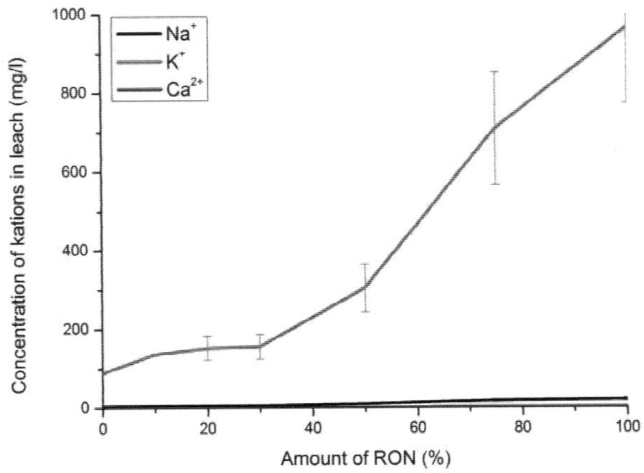


b)

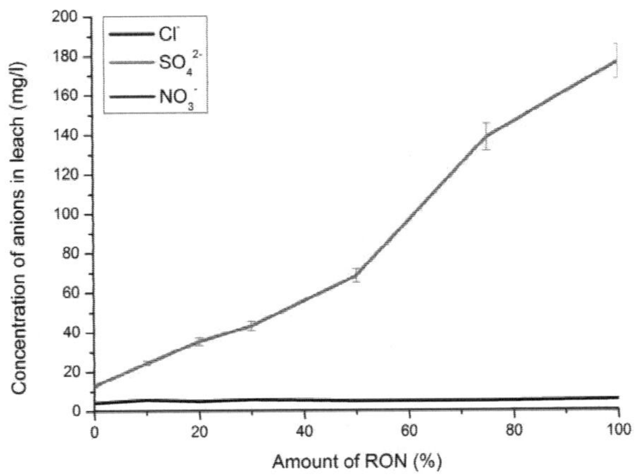


**Figure 4.** Development of concentration of cations (a) and anions (b) in leachate in dependence on RON content in the geopolymer matrix.

## 4. Conclusions

For the preparation of geopolymer materials, the flue dust of 0-100% was used as a substitute in blast-furnace slag and metakaolin matrix. It was found that this material can also be used successfully for the preparation of geopolymers, but due to the content of kaolinite, which dissolves very slowly in the alkaline environment, such a high-quality and appropriate aluminate-silicate structure is not formed. The addition of the flue dust results in reduced compressive and flexural strengths, prolongation of setting time and affects the flowability of the mixtures.

**Acknowledgments**

The authors wish to express their gratitude and sincere appreciation to the authority of Ministry of Industry and Trade of the Czech Republic, project No. FV30150 for financial support.

**References**

[1] Davidovits J 2011 *Geopolymer Chemistry and Applications* (France, Institute Geopolymere) p 623

[2] Van Jaarsveld J G S, van Devener J S J and Lukey G C 2002 The effect of composition and temperature on the properties of fly ash- and kaolinite-based geopolymers *Chemical Engineering Journal* **89** pp 63–73

[3] Bortnovsky O, Sobalik Z, Tvarůžková Z, Dědeček J, Roubíček P, Průdková Ž and Svoboda M 2005 Structure and stability of geopolymers synthesized from kaolinitic and shale clay residues *Proceeding of the world congress geopolymer* pp 81–84

[4] Longhi M A, Rodríguez E D, Bernal S A and Provis J L 2016 Valorisation of a kaolin mining waste for the production of geopolymers *J. Clean. Prod.* **115** pp 265–272

[5] Aboulayt A, Riahi M, Ouazzani T, Hannache H, Gomina M and Moussa R 2017 Properties of metakaolin based geopolymer incorporating calcium carbonate *Advanced Powder Technology* **28** pp 2393–2401

[6] Komnitsas K and Zaharaki D 2007 Geopolymerisation: A review and prospects for the minerals industry *Miner. Eng.* **20** pp 1261–1277

[7] Heah C Y, Kamarudin H, Mustafa Al Bakri A M, Bnhussain M, Luqman M and Khairul Nizar I 2012 Study on solids-to-liquid and alkaline activator ratios on kaolin-based geopolymers *Constr. Build. Mater.* **35** pp 912–922

[8] Xu H and Van Deventer J S J 2002 Geopolymerisation of multiple minerals *Miner. Eng.* **15** pp 1131–1139

[9] Firdous R, Stephan D and Djobo J N Y 2018 Natural pozzolan based geopolymers: A review on mechanical, microstructural and durability characteristics *Constr. Build. Mater.* **190** pp 1251–1263

[10] Pacheto-Torgal F, Labrincha J A, Leonelli C, Palomo A and Chindraprasirt P 2015 *Handbook of Alkali-activated Cements, Mortars and Concretes* (United Kingdom, Elselvier) p 830

[11] Tchakoute H K, Elimbi A and Yanne E 2013 Utilization of volcanic ashes for the production of geopolymers cured at ambient temperature *Cement Concrete Comp.* **38** pp 75–81

[12] Chindaprasirt P and Rattanasak U 2010 Utilization of blended fluidized bed combustion (FBC) ash and pulverized coal combustion (PCC) fly ash in geopolymer *Waste Manage.* **30** pp 667–672

[13] Ryang H M and Sim H C 2019 Formation of CaCO3 varieties from a carbonated aqueous solution *Carbonate. Evaporite.* **34** pp 1619–1626

IOP Conf. Series: Materials Science and Engineering 1039 (2021) 012016 doi:10.1088/1757-899X/1039/1/012016

# Mineralization of wooden elements by bypass cement kiln dust

**T Majstrikova[1], F Khestl[1], P Mec[1] and B Vojvodikova[1]**

[1] Faculty of Civil Engineering, VSB - Technical University of Ostrava, Ludvika Podeste 1875/17, 708 00 Ostrava-Poruba, Czech Republic

Email: tereza.majstrikova@vsb.cz

**Abstract**. The application of renewable and waste materials for the production of building materials is currently one of the main trends, especially with regard to environmental aspects. This is a reduction in the volume of materials that could be a waste and at the same time a reduction in the cost (including carbon footprint) for special wood treatment products that would have to be produced. One example given in this article is the possible application of a mineralization of wood matrix by kiln dust produced in a cement plant. This treatment can be primarily used in the production of wood-based panels bonded with cement, whose main purpose is to limit the leaching of soluble components, which affect the setting of the cement. If the cement kiln dust is applied for the mineralization of wooden elements, the flammability of wood can be positively influenced. Bypass cement kiln dusts contain significant amount of chloride, which in this case function as a reducing agent. In this article, the effect of the wood mineralization by the cement kiln dust from CEMMAC a.s. Horné Srnie was investigated on thermal degradation of wood by a simultaneous thermal analysis TG / DSC.

## 1. Introduction

Wood is an organic renewable material that has been and still is preferably used for many construction purposes like ceiling and roof constructions, lintels, claddings, floors, windows, doors, stairway construction, vertical load-bearing structures etc. This wide spectrum of wood application in civil engineering was caused mainly by its low bulk density, good workability, strength, thermal insulation parameters and also by other properties [1]. However, this material is also characterized by some unsuitable properties such as reduced durability and flammability. Various types of treatment products are applied to reduce combustion and promote the formation of a protective carbonized layer [1–2]. On the other hand, the flammability can be also reduced by applying a certain production technology, as in the case of wood-cement panels [3].

Flammability of wooden elements has a direct connection with the amount of ash (non-combustible fraction) and the release of heat during combustion [1–2]. These parameters can be influenced by the application of bypass cement kiln dust (BCKD). Firstly, increasing the non-combustible proportion is determined by the principle of wood mineralization, i.e. a natural phenomenon, which is generally based on the deposition of mineral substances that can partially or completely replace organic substances [4–6]. Conversely, substances acting as oxidizing agents can be used to reduce the released heat especially with larger wood particles. These substances include chlorine, which is a strong oxidizing agent - it reduces itself (increases the oxidation number), thereby reducing the amount of oxygen involved in the reaction during the combustion process (figure 1) [1–2]. At the same time, chloride-based compounds are also used as accelerating agents that technologically affect the setting time and the onset of strength in the early stages in wood-based composites [3].

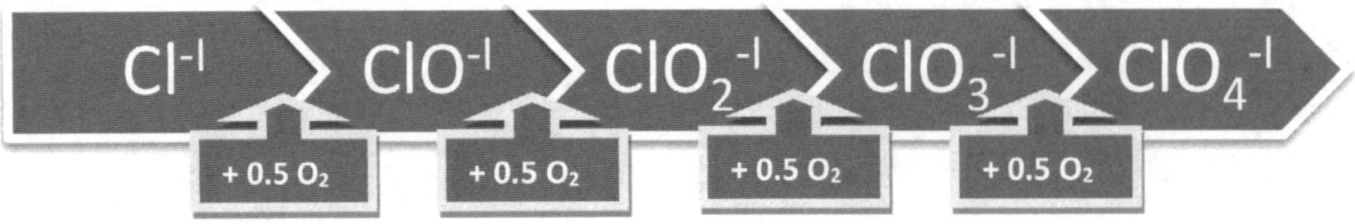


**Figure 1.** Gradual chlorine oxidation.

Thermal analysis is an effective analytical tool for this study because wood with a high content of combustible components is relatively easily decomposed at elevated temperature [1–2, 4]. The basic output is the exothermic and endothermic phenomena observed on the DSC curve and the weight changes detected on the TG curve [1–2, 4, 7]. The results are possible to compare with the thermal curves database in some cases [8] or process in the form of quantitative analysis [7]. Thermal analysis is often used for assessment of basic wood components [9] or targeted treatments with the task of monitoring changes in thermal degradation of material, primarily seen in the changes of mass residue in TG curve, temperature shifts of the thermal decomposition of wood components in DSC curve [2, 4]. For this reason, the characteristics of wood flammability for the reference (untreated) samples and samples treated with a solution of BCKD were determined in the experimental part by using of thermal analysis. It is essential to determine the non-combustible fraction and the shift of temperature milestones of wood thermal degradation by verified and published methodology [2, 4].

## 2. Materials and Methods

### *2.1. Wood*

Primarily, specimens for the experiment 20 × 20 × 30 mm (R × T × L) were prepared from one representative of softwood species - Norway spruce (*Picea abies L.*) and one representative of hardwood species - European beech (*Fagus sylvatica L.*). The specimens were placed in a laboratory environment at the temperature of 20 ± 1°C and the relative humidity of 50 ± 5%. Every specimen was weighted with an accuracy of 0.1 g and measured with the accuracy 0.01 mm.

### *2.2. Chemicals*

For the testing, the BCKD from the CEMMAC a.s. Horné Srnie production plant was used.

**Table 1.** Chemical composition [wt. %] of the bypass cement kiln dust (BCKD) determined by XRF.

| Chemical component | Portion [wt. %] |
|---|---|
| $SiO_2$ | 5.32 |
| $Al_2O_3$ | 1.46 |
| $Fe_2O_3$ | 1.22 |
| MgO | 0.48 |
| CaO | 34.08 |
| $Na_2O$ | 0.31 |
| $K_2O$ | 18.62 |
| $TiO_2$ | 0.12 |
| $SO_3$ | 5.16 |
| $P_2O_5$ | 0.05 |
| Cl | 10.49 |
| Br | 0.16 |
| MnO | 0.09 |

The chemical composition of the used BCKD can vary slightly. On the other hand, approximate chemical composition determined by XRF can be found in table 1. The treatment solution was prepared by mixing the powder sample from BCKD with demineralized water in a ratio of 5, 10, 15 or 25 g of the BCKD to 100 ml of demineralized water. The pH of each solution was measured.

### *2.3. Preparing of specimens*

Wooden specimens were treated with BCKD water solution by a coating or by a dipping method at the atmospheric pressure of ≅ 100 kPa, the temperature of 20 ± 1°C, and with different immersion time – 30, 60, 180 minutes. The specimens were surface dried with filter paper after removing from the solution. Every specimen was weighted after treatment with an accuracy of 0.1 g. The retention parameters R [g·m$^{-2}$] were calculated by (1) from the weight after immersion $w_{after\text{-}treat}$ in relation to the weight before treatment $w_{before\text{-}treat}$ and the specimen surface S [mm$^2$].

$$R = 1000000 \times \frac{w_{after-treat} - w_{before-treat}}{S} \quad (1)$$

Subsequently, these specimens were conditioned in a laboratory environment at the temperature of 20 ± 1°C and the relative humidity of 50 ± 5% until they reached the equilibrium moisture. Reference specimens (without mineral treatment) were also conditioned. For the thermal analysis, superficial cuts in the form of chips were taken by a scalpel. The reference specimens were also taken by a scalpel for thermal analysis, similar to the treated samples.

### *2.4. Thermal analysis*

Thermal analysis was performed by SDT Q600, TA Instruments, which allows a simultaneous combination of thermogravimetric analysis and differential scanning calorimetry (TG / DSC) [10]. The wooden superficial cuts were placed in corundum pans and heated up at a constant heating rate of 10°C·min$^{-1}$ to 700°C in an oxidizing atmosphere of air. The oxidizing atmosphere allows simulation of fire conditions that could confirm the fire retardation effects.

An important output was the shift of values of non-combustible fraction generated after thermal degradation of predominantly volatile and solid components (TG curve), which may indicate a change in material flammability [2, 4]. The shift in the initial temperature (the beginning of hemicellulose decomposition) should be also observed by the TG curve. The evaluation of the DSC curve was added for temperature definition of the peak for thermal degradation of predominantly volatile components and the first peak for the decomposition of solid components [2, 4].

## 3. Results and Discussion

### *3.1. The pH measurement*

The pH was measured for all solutions, which was made of 5, 10, 15, or 25 g powder samples from the BCKD and 100 ml demineralized water. A relatively high pH of 13.5 was already measured for the solution made from 5g of the BCKD and 100 ml of demineralized water. For all other solutions, the pH values have already been at a maximum of 14. This phenomenon may be advantageous for wood, alkaline nature of the solution can also reduce the development of molds and fungi on this material.

### *3.2. Weight changes of specimens*

For the treated specimens, the retention parameter R immediately after treatment was determined in relation to the surface of the specimen which is in contact with the BCKD solution (table 2 and 3).

Based on the results it can be stated that the weight of the specimens increases with the increasing exposition time, particularly due to the solution intake and the deposition of the mineral particles. At the same time, it is apparent that solutions of lower concentrations penetrate the wood considerably better [5]. Especially for Norway spruce, it is apparent that the uptake of the solution at the concentration of 10 g of the BCKD per 100 ml demineralized water is significant and a considerable mineral particle deposition can also be expected.

**Table 2.** Retention parameter R for Norway spruce specimens.

| | R [g·m$^{-2}$] | | | |
|---|---|---|---|---|
| | *Concentration 5 g per 100 ml* | *Concentration 10 g per 100 ml* | *Concentration 15 g per 100 ml* | *Concentration 25 g per 100 ml* |
| Coating | 210.6 | 127.9 | 104.4 | 206.6 |
| Immersion 30 minutes | 566.6 | 915.8 | 730.8 | 520.5 |
| Immersion 60 minutes | 523.3 | 546.5 | 825.4 | 845.7 |
| Immersion 180 minutes | 662.6 | 1034.7 | 1036.6 | 895.4 |

**Table 3.** Retention parameter R for European beech specimens.

| | R [g·m$^{-2}$] | | | |
|---|---|---|---|---|
| | *Concentration 5 g per 100 ml* | *Concentration 10 g per 100 ml* | *Concentration 15 g per 100 ml* | *Concentration 25 g per 100 ml* |
| Coating | 114.6 | 118.2 | 130.3 | 153.6 |
| Immersion 30 minutes | 930.9 | 922.6 | 445.6 | 599.6 |
| Immersion 60 minutes | 994.3 | 547.1 | 414.6 | 611.6 |
| Immersion 180 minutes | 1559.4 | 900.7 | 725.1 | 744.2 |

### *3.3. Thermal analysis.*

The values of non-combustible fraction for the treated specimens are stated in tables 4 and 5. These values can be compared with the reference specimens, for which mass residue take value 0.00% for Norway spruce and 0.20% for European beech.

**Table 4.** Non-combustible fraction after thermal decomposition for Norway spruce specimens.

| | Non-combustible fraction [%] | | | |
|---|---|---|---|---|
| | *Concentration 5 g per 100 ml* | *Concentration 10 g per 100 ml* | *Concentration 15 g per 100 ml* | *Concentration 25 g per 100 ml* |
| Coating | 0.00 | 0.00 | 0.61 | 3.24 |
| Immersion 30 minutes | 2.40 | 4.16 | 5.15 | 6.52 |
| Immersion 60 minutes | 2.11 | 7.62 | 4.07 | 7.42 |
| Immersion 180 minutes | 2.40 | 10.32 | 8.46 | 6.87 |

Based on the test results, it can be stated that the non-combustible fraction values increase in most cases with prolonged exposure to the BCKD solution. In terms of solution concentrations, it can be emphasized that too high a concentration of the BCKD solution may not correspond to higher values of the non-combustible fraction, as the retention parameters already showed. However, too low concentrations do not guarantee uniform mineralization. For the above reasons, the concentration of 10 g of the BCKD per 100 ml of demineralized water appears perspective.

**Table 5.** Non-combustible fraction after thermal decomposition for European beech specimens.

| | Non-combustible fraction [%] | | | |
|---|---|---|---|---|
| | *Concentration 5 g per 100 ml* | *Concentration 10 g per 100 ml* | *Concentration 15 g per 100 ml* | *Concentration 25 g per 100 ml* |
| Coating | 0.00 | 0.99 | 0.00 | 2.75 |
| Immersion 30 minutes | 5.42 | 5.17 | 3.40 | 4.53 |
| Immersion 60 minutes | 2.96 | 5.21 | 3.48 | 7.64 |
| Immersion 180 minutes | 1.02 | 5.13 | 6.36 | 8.15 |

Thermal analyses of specimens with different exposition in the BCKD solution are shown in figure 2–5. The various figures show the TG curves – monitoring the weight changes in % and the DSC curves – describing heat flow in °C·mg$^{-1}$. Figure 2 also defines the location of key characteristics for thermal degradation of wood, namely non-combustible fractions, peaks of predominantly volatile components, first peaks of solid components and initial temperatures.

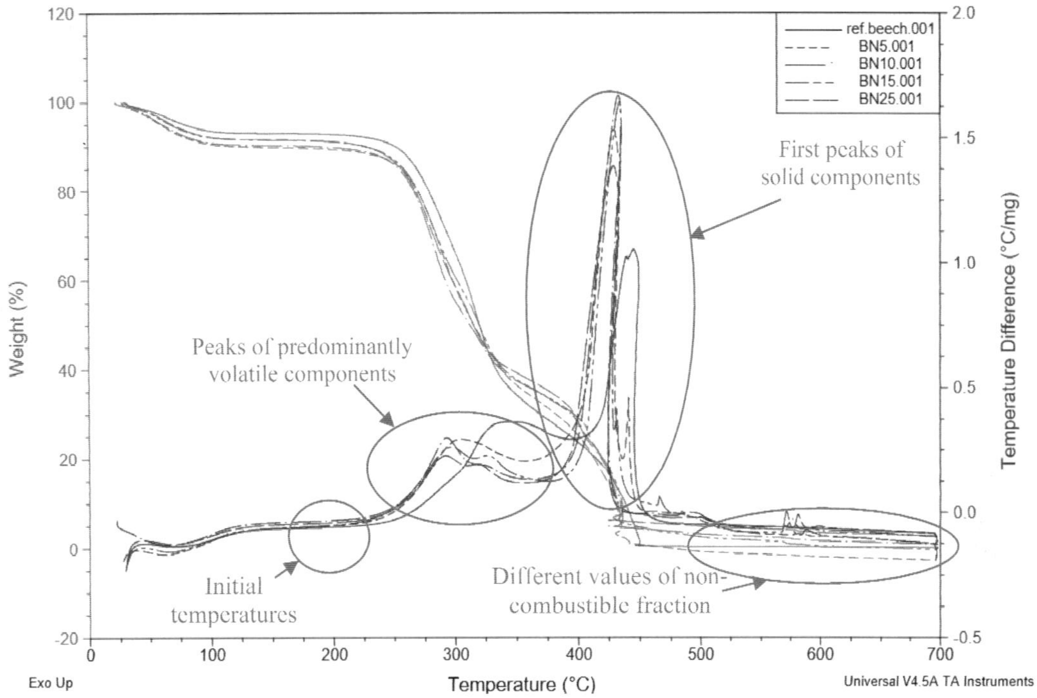


**Figure 2.** Comparison of thermal analysis curves of beech surface chips: reference specimen (ref_beech), treated specimens by coating with solution of 5 g BCKD and 100 ml demineralized water (BN5), with the solution of 10 g BCKD and 100 ml demineralized water (BN10), with the solution of 15 g BCKD and 100 ml demineralized water (BN15) or with the solution of 25 g BCKD and 100 ml demineralized water (BN25).

It is apparent from the graphical records of analysis that low concentration of the solution or short-term exposure in the solution, for example by coating or spraying, does not significantly affect the course of combustion (figure 2).

The opposite is the case with treatment by dipping method (figure 3) or higher concentrations of the BCKD in solutions (figure 4), where the changes in peak size are evident in the DSC curves. It can be stated that increasing intake of the solution results in a smaller peak area thus, less heat is generated. Also, the first peaks of the solids components have a slight shift to higher temperatures, which may also indicate a change inflammability of wood. It is also apparent from the TG curve that the weight changes for these specimens are considerably more gradual.

On the other hand, thermal analyses show new peaks from 450°C, which are hardly found in smaller amount of the BCKD and coating applications. These peaks are always supplemented by weight changes; therefore, some decomposition of the specimen components is a probable cause. Moreover, the peaks do not appear in the case of thermal analysis in the reduction atmosphere. All the results show that the cause may be just high chloride content, which would also confirm the partial retardation effect. However, these reactions cannot be specified without further chemical analyses.

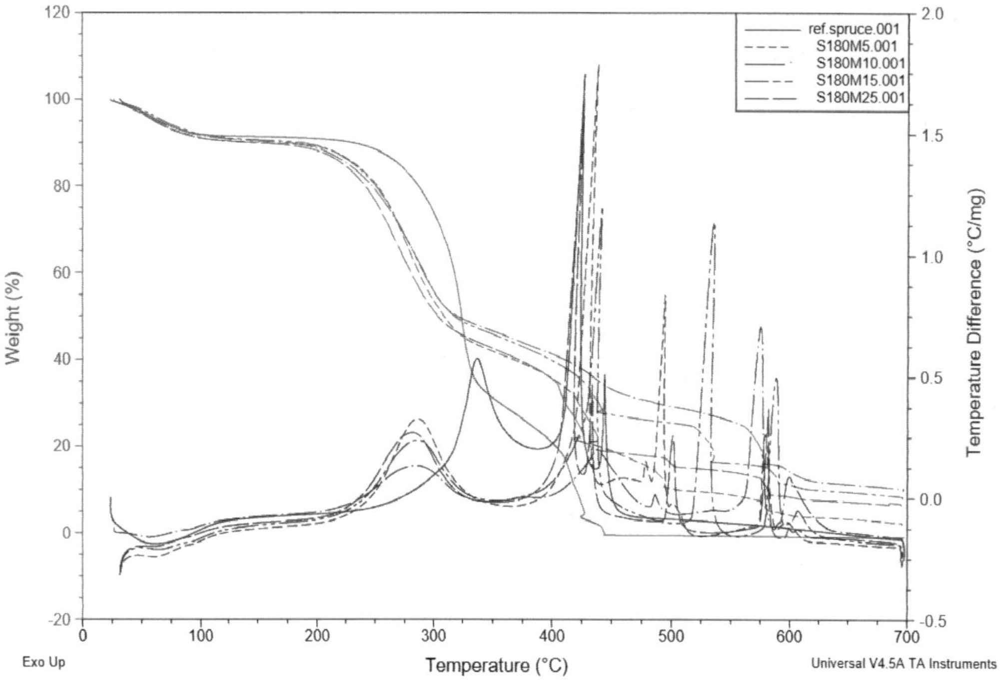


**Figure 3.** Comparison of thermal analysis curves of spruce surface chips: reference specimen (ref_spruce), treated specimens by dipping method with immersion time 180 minutes and with the solution of 5 g BCKD and 100 ml demineralized water (S180M5), with the solution of 10 g BCKD and 100 ml demineralized water (S180M10), with the solution of 15 g BCKD and 100 ml demineralized water (S180M15) or with the solution of 25 g BCKD and 100 ml demineralized water (S180M25).

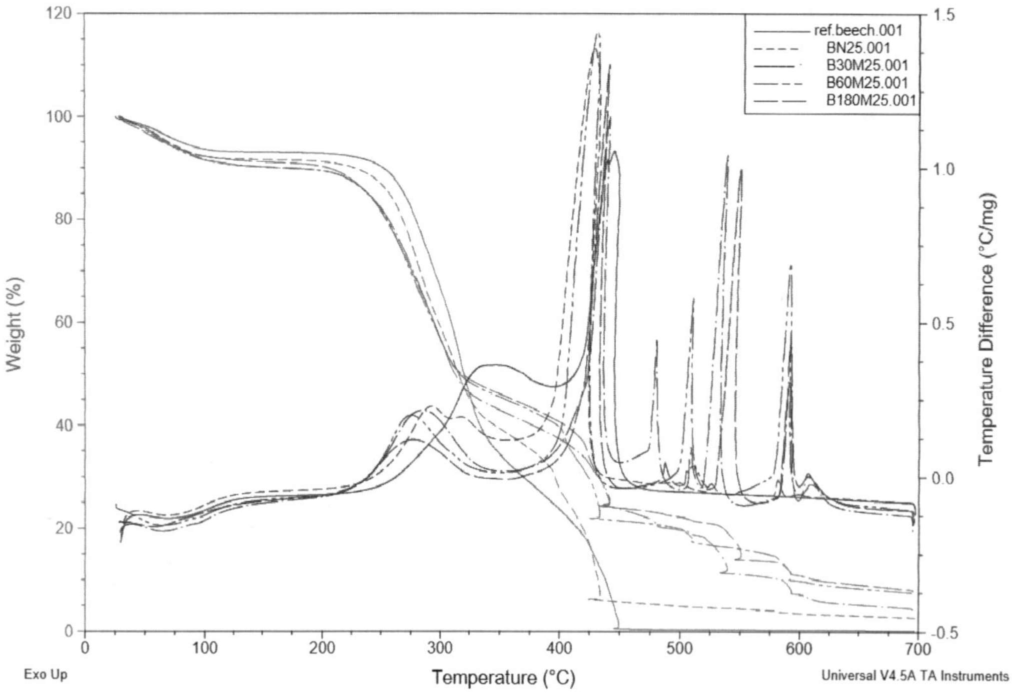


**Figure 4.** Comparison of thermal analysis curves of beech surface chips: reference specimen (ref_beech), treated specimens with the solution of 25 g BCKD and 100 ml demineralized water by

coating (BN25), or by dipping method with immersion time 30 minutes (B30M25), 60 minutes (B60M25) or 180 minutes (B180M25).

A concentration of 10 g of the BCKD per 100 ml of demineralized water for Norway spruce can be seen in figure 5. This concentration appeared to be very effective in terms of non-combustible fraction growth. From the heat flow point of view, it can be seen that for all treatments the peak size of predominantly volatile components decreased. Peaks of solid components are on the contrary narrower compared to the reference specimens. The reason for these reactions is the application of the BCKD with a higher amount of chlorines. The above-mentioned reactions above 450°C occur only for the specimens with higher immersion time.

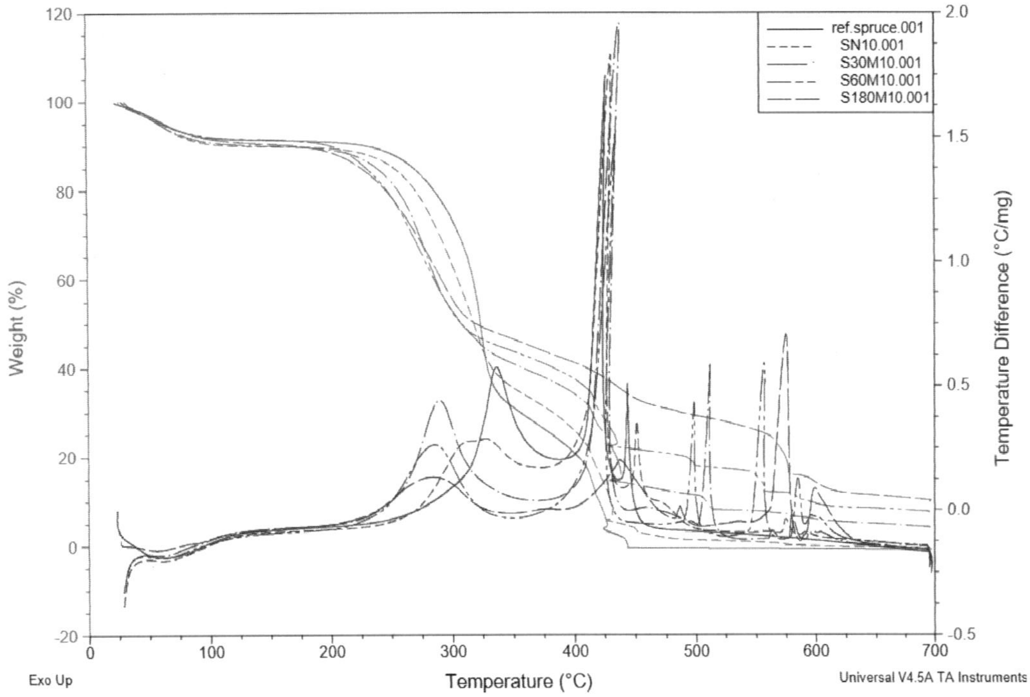


**Figure 5.** Comparison of thermal analysis curves of spruce surface chips: reference specimen (ref_spruce), treated specimens with solution of 10 g BCKD and 100 ml demineralized water by coating (SN10), or by dipping method with immersion time 30 minutes (S30M10), 60 minutes (S60M10) or 180 minutes (S180M10).

It has been shown that the application of BCKD-based treatments can reduce the flammability of the treated wood. However, prolonged exposure to the solution is required and therefore application in the form of a coating or a spraying is not very suitable. If we observe a parameter associated with flammability, a specifically the non-combustible fraction, higher values can be achieved by a longer immersion time. Furthermore, it has been found that lower concentrations of the solution penetrate much more easily into the specimens. This effect is well known from the petrification process, wherein a faster saturation of the organic matrix with minerals is achieved in solutions with lower concentrations [5, 11]. On the result basis it can be stated that the concentration of 10 g of the BCKD per 100 ml of demineralized water can be optimal for this purpose.

Based on the DSC curves, the thermal degradation of specimens with higher solution retention is limited, especially in the area of thermal decomposition of the basic wood components. The primary cause can be the known effect of chlorines as a fire retardant [1]. However, the effect of the individual chloride salts is usually investigated [2]. In our case, chloride salts, in particular KCl, are also present in the BCKD. On the other hand, the BCKD also contains a number of other substances that can affect the combustion process [12]. In particular, silicates may form layers on the inner or the outer surface which can partly protect the organic substance against fire, but it may partially decompose during heating [1, 8]. It can be assumed that the resolution of these effects is not fully possible due to the complexity of the problems.

## 4. Conclusion

The high chloride content limits the use of a bypass cement kiln dust (BCKD) for conventional cement applications. On the other hand, chloride-containing substances are used for some applications related to the application of renewable natural materials. The application of chloride-based substances is frequent in the case of accelerated setting time of the wood particleboard bonded with cement. Chloride salts are also used as fire retardants for timber elements.

The fire-retarding effect of BCKD has been studied by thermal analysis in this experiment. It has been shown that application of BCKD-based treatments can reduce the flammability of the treated wood. The reason for the reduction in flammability is mainly the content of chloride salts and partly silicates in BCKD. On the other hand, prolonged exposure to the solution and lower concentration of the water solution is required. For this case, the concentration of 10 g of the BCKD per 100 ml of demineralized water and the immersion in the order of hours can be recommended. However, further analytical testing should be performed to confirm these conclusions.

For this reason, the treatment method with BCKD water solution can be recommended mainly for the preparation of wood-based composites and potentially for solid wood application.

**Acknowledgement**

This article was prepared thanks to project Geodust. This project has received funding from the European Union's Horizon 2020 research and innovation programme under grant agreement No. 734833.

## References

[1] Reinprecht L 2004 *Ochrana dreva* (Zvolen: Technická univerzita vo Zvolene) p 453
[2] Tang W K and H W 1968 *Effect of Inorganic Salts on Pyrolysis of Wood, Cellulose, and Lignin Determined by Differential Thermal Analysis* (Madison: Forest Products Laboratory) p 30
[3] Matoski A, Hara M M, Iwakiri S and Machado Casali J 2013 *Acta Scientiarum technolgy* **35** pp 655–660
[4] Majstríková T, Mec P and Sviderová E 2019 *Key Engineering Materials* **808** pp 60–65
[5] Akahane H, Furuno T, Miyajima H, Yoshikawa T and Yamamoto S 2004 *Sedimentary Geology* **169** pp 219–228
[6] Kim K W, Yoon Ch J, Kim P G, Lee M B and Lim J H 2009 *Micron* **40** pp 519–525
[7] Šesták J 1982 *Měření termofyzikálních vlastností pevných látek: Teoretická termická analýza* (Praha: Academia) p 348
[8] Ramachandran V S, Paroli M R, Beaudoin J J and Delgado H A 2002 *Handbook of thermal analysis of construction materials* (Norwich: William Andrew) p 680
[9] Brebu M and Vasile C 2010 review *Cellulose chemistry and technology* **44** pp 353–363
[10] TA Instruments - Thermal Analysis *TA Instruments* 2010 (Online: http://www.tainstruments.com/pdf/brochure/sdt.pdf)
[11] Buurman P 1972 *Scripta Geol.* **12** pp 1–43
[12] Peethamparan S, Olek J and Lovell J 2008 *Cement and Concrete Research* **38** pp 803–815

IOP Conf. Series: Materials Science and Engineering 1039 (2021) 012017 doi:10.1088/1757-899X/1039/1/012017

# Energy-active roof system – design, development and testing

**P Bibora**[1], **R Cechmanek**[1], **P Leber**[2] **and L Vlach**[3]

[1] Research Institute for Building Materials, Hnevkovskeho 30/65, 617 00 Brno, Czech Republic
[2] DAKO Brno, spol. s r.o., Prace, Czech Republic
[3] SOLUTRON, s.r.o., Brno, Czech Republic

Email: bibora@vustah.cz

**Abstract**. The aim of the research is a development of a new energy effective roof system, fulfilling a function of a roof covering as well as solar active element, reducing energy performance of buildings. There were designed and tested energy active roof elements, consisted of glass-fibre-reinforced-concrete carrier with integrated photovoltaic and heat-exchange layer, which ensures cooling of photovoltaic elements as well as enables a transfer of excessive thermal energy into a storage tank for a later use. The effort of project solvers is to achieve entirely new economic efficient using solar energy by means of energy active roof covering. Current research and tests proved, that there is a possibility to create such a system using present materials and technological solutions. However, for the successful realization it is necessary to solve further technological aspects, namely mutual connection of the roof elements and prove load-bearing capacity and fire resistance of the designed system.

## 1. Introduction

Presented article briefly describes some of the phases of a research project focused on development and testing of a complete roof system that would serve as both cladding system and solar energy accumulator. The most appropriate materials for such purposes seem to be mutually connected thin-walled glass-fibre concrete (GFC) parts with integrated photovoltaic (PV) and heat-transfer (HT) layer. Such roof design would lower the energy demands of the building and fully substitute a standard roof system at the same time. Additionally, the excessive heat generated by PV cells would be transferred by liquid in a TC layer directly into a heat exchanger for further use in the building. Such cooling system would lead to higher energy yields of the PV cells and simultaneously prevent overheating inside the building. Mutual connection of the PV and HE systems must be solved so that an integrated roof cladding system is created. [1]

Thin-walled composites made from fine-grain cement-sand matrix reinforced by alkali-resistant glass fibres are currently used in numerous applications in building industry, mainly as architectural elements and facade panels. Their advantages are low weight and sufficient durability against mechanical stress and weather. These prerequisites led to the idea of this innovative design of roof tiles that produce both electricity and thermal energy. Solar energy will be transformed into electric and thermal energy thanks to an integrated system of photovoltaic cells and working fluid tubing on a glass-fibre carrier. The heat acquired will be stored via heat exchanger in a hot water storage tank. In comparison with typical photovoltaic panels, the roof cladding, the solar panels and the heat collectors are integrated in one complex functional unit.

A typical solar collector is composed of a tubular system that absorbs solar rays and transform them into thermal energy. The heat is prevented from discharging into the air and is transported by working fluid into the heat exchanger which serves also as a storage tank. Nowadays, commercially available solar collectors are products of an advanced development, however, high initial costs and susceptibility to damage during transport and assembly are still their major disadvantages. [1]

The subject of this project is design and development of solar-active roof tiles whose production and assembly would be cheaper while maintaining the same efficiency and life span as the standard solar panels. Rich experience of both participants were used for the development of the mixing GFRC formulas. DAKO Brno company deals with the production and installation of ventilated facades of fiberglass concrete and we also used the experience mainly from research projects dealing with the development of high-strength fiber-reinforced concrete [2, 3].

Suggested solution using glass-fibre carrier for installation of working fluid tubes and photovoltaic cells is an innovative approach that also fulfils a construction function as the resultant roof tiles can fully replace a typical roof cladding. [1]

## 2. Technical and material solution of the roof system

The aim of the project is to design and validate the whole new roof cladding system for sloped roofs of residential and industrial buildings, consisting of glass-fibre reinforced concrete roof tiles with integrated photovoltaic and heat-transfer layers.

### *2.1. Material composition*

There were set mixing formulas for production of glass-fibre-reinforced-concrete (GFRC) carriers and subsequently tested their physico-mechanical parameters. At the same time, the most suitable types of photovoltaic panels were selected on the base of testing their energy efficiency and material characteristics.

The glass-fibre reinforced concrete (GFRC) element is made of fine-grain cement-sand matrix supplied with chemical admixtures, especially additives modifying texture or hydrofobization agents. The basic matrix is reinforced with spread glass-fibres and sprayed into moulds, sometimes spraying is combined with pouring (premix). The filler:binder:reinforcement ratio is determined by the physico-mechanical properties required, mainly flexural and impact strength values. The usual dose of binder is 7–65% (typically 50%), dose of filler is 10–70% (typically 50%) and dose of fibre 0.5–10% (typically 5%). The composition of the mixture guarantees stability of physico-mechanical properties for 30 years in common climate/weather conditions. Figure 1 shows an implementation example of a developed roof system.

**Figure 1.** Energy-active roof panels – implementation example.

### *2.2. Shape and system solution*

The technical details of the manufacture procedure of the proposed GFC panels with integrated PV and HT layer were solved in this phase of research. The HT layer has been designed to ensure efficient transfer of thermal energy from the PV module to a carrier liquid medium with sufficiently low hydraulic resistance, so that individual GFC elements can be serially chained according to the required parameters.

Therefore, it was necessary to design and verify the main key details of the HT layer in order to ensure the optimal transfer of thermal energy. After evaluating the results, a type of structural element in the form of a scalloped flexible steel tube with a diameter of 11.7 mm and a wall thickness of 0.26 mm was selected, with respect to the properties and thermal conductivity of the GFC panel in which the HT layer is intercalated.

**Figure 2**. Possible layout of developed HT layer.

In relation to optimal HT layer efficiency, the placement of the HT layer structural elements was designed and verified by measuring of the resulting roof system elements. The measurement was also supplemented with results from the software module calculation performed on the VUT FEKT in Brno. The results of the measurements showed the optimum distribution of the HT structural element pipes at a distance of 50–75 mm from each other. One example of possible layout of developed HT layer is shown in figure 2.

During designing of production technology of GFC elements, limiting values of production and assembly technology were taken into account. An important factor influencing the final measures of the element (together with photovoltaic (PV) and heat-transfer (HT) layers, is its weight. Given that the complete roof system must be composed of several different sizes of elements, smaller elements are designed individually. Elements intended to complete the functional composition are atypical, with neither the PV nor the HT layer.

Overall, two variants of the constructional-technical solution of the system, which differ in the manner of anchoring, were verified. The first variant, which is expected to be for new buildings, is anchored as a larger roof element by means of anchors attached through the roof batten to the rafters. The second variant, designed mainly for reconstructions, is anchored similarly, but only on additionally installed vertical roof battens.

Both variants of the constructional-technical solution of the system also differ in their technical parameters, ie different volt-amp characteristics and different heat recovery properties from the HT layer. However, the performance per 1 $m^2$ is similar for both variants.

*2.3. Solution of electrical interconnection of individual PV roof elements*

In our measuring kits, we used only 1 + 1 PV modules to compare current data. The panels tested had a $U_{oc}$ voltage of 12 $<U_{oc}<$20V. This voltage is low for further commercial use, it will be necessary to connect these modules into strings series in a real installation, to get higher voltage as well as higher power of the whole assembly for further processing. In terms of electrical connection of the final modules, we will be limited only by the maximum voltage given by the manufacturer of the PV layer (1000 V) and by the voltage required for the operation of the next link in the energy processing chain - inverters. However, it is necessary to assess the individual strings with a heat transfer layer in terms of the limitations associated with the maximum liquid flow rate through the heat transfer layer of the individual PV modules.

For further utilization of produced DC energy, it is possible to consider:

- off grid systems
- modern hybrid systems with HV accumulator
- hybrid systems with low voltage accumulators
- on-grid connection
- on-grid connection in combination with optimizers and inverter

On-grid connection with optimizers and inverter is best suited for our purpose, see Figure 3. With this type of connection we can combine not only modules with different performance, but also modules with and without heat exchange layer and without loss of performance. Optimizer manufacturer software will allow the designer to lay out roof panels for a specific application. The designer will be able to select the inverter and the optimizer and connect them. In the next step, the software will calculate the parameters of installation and future production. This eliminates the possibility of errors or poor design choices. After successful completion of the design, the project can be turned into monitoring and the whole system can be monitored via the corporate cloud. The entire system can therefore be perfectly supervised and evaluated.

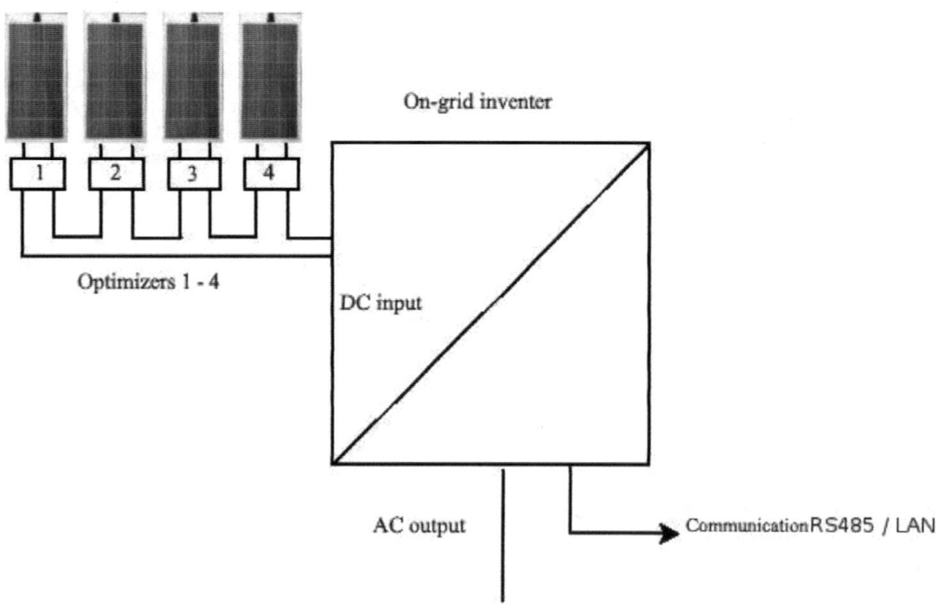


**Figure 3**. On-grid connection in combination with optimizers and inverter.

It is possible to use another system, but it is always necessary to carefully check the individual parameters during the design, especially the minimum DC voltage of the inverter.

## 3. Experimental works

### *3.1. Measurement of GFC panels with integrated PV and HT layers*

The aim of the measurement was to verify the heat transfer efficiency from the integrated PV layer to the HT layer in the GFC panel. The measurement was carried out at 3 measuring stations with regard to the environment, exposure and diversity of used roof elements. Two sites were selected in the urban agglomeration (approx. 200 above sea level) and one measurement in an environment with a higher altitude of approx. 500 above sea level.

The measurements showed slightly different values, which were caused by different meteorological conditions. For the evaluation we considered only the period from April to the end of September. The same pair of roof elements connected individually was used for the measurements, with the one test specimen without active TS layer and one test specimen with integrated active TS layer. The whole measuring system can be safely used in the seasons just from April to September. After this period, the heat obtained has a lower nominal temperature and is therefore unsuitable for direct storage. There is also a risk of frost damage to the measuring system. It can be stated that the roof elements continue to generate a temperature difference of 1.5 to 2.5°C during sunny days. This predetermines the system utilization, for example, as the primary heat source for the heat pump, in the case the antifreeze heat transfer medium will be used.

The measurements showed that the roof elements with active heat transfer layer had about 8% higher performance due to heat removal. Moreover, the thermal energy generated was directly proportional to the electrical energy obtained. The daily yield was in the range 1,100 to 2,000 J.

### *3.2. Methodological procedure for energy efficiency measuring*

The energy efficiency of the developed roof elements was measured in terms of proportion. Two nearly identical PV modules, in terms of dimensions and performance, were compared. This was achieved by laboratory measuring of the I/V characteristics of PV modules on the PASAN Sun Sim3C tester. The selected PV modules were applied to the manufactured GFC panels with and without integrated HT layer and also another measurement was made to see whether the PV layer was damaged during application.

As shown in the graphs in figures 4 and 5, we were able to select pairs with very similar I/V characteristics and with a power difference $P_{mpp}<0.5W$ at its operating point.

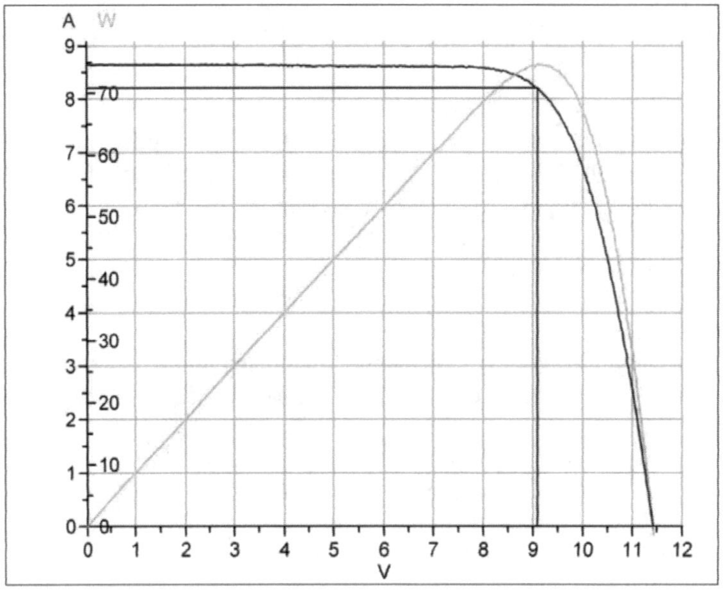


**Figure 4.** I/V characteristic of PV module without active HT layer.

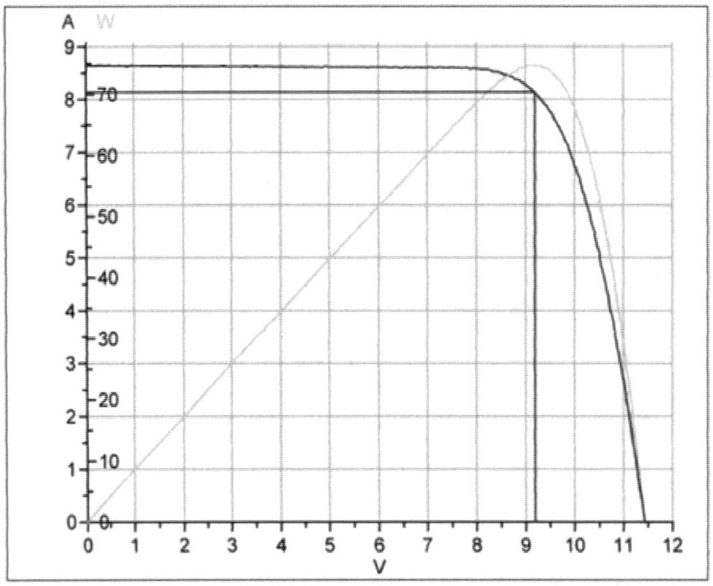


**Figure 5.** I/V characteristic of PV module with active HT layer.

The basic measurement took place over time and the electrical power of the PV panel and the heat extracted from the HT layer were monitored in dependence on the incident sun radiation. The realized measurements showed that the change in power depending on the incident radiation was rapid, whereas the value of the removed heat changed slowly due to the accumulation ability of the GFC elements.

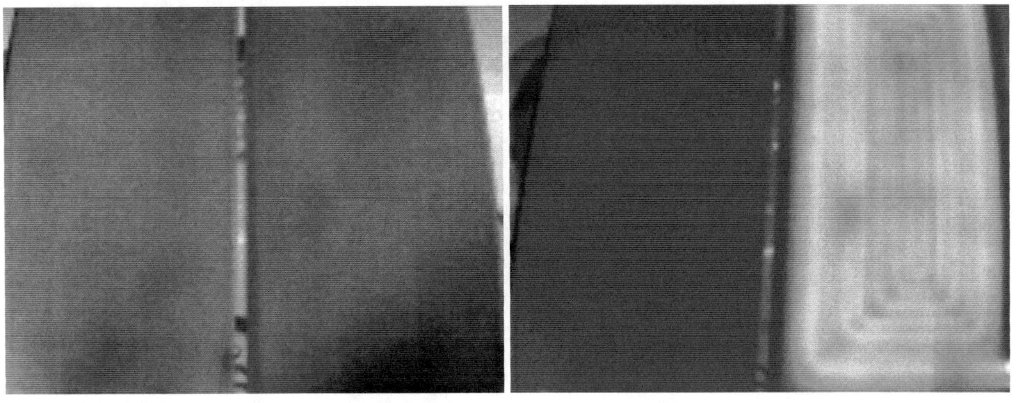

**Figure 6.** Temperatures of GFC elements without active TS layer (left) and with active TS layer (right).

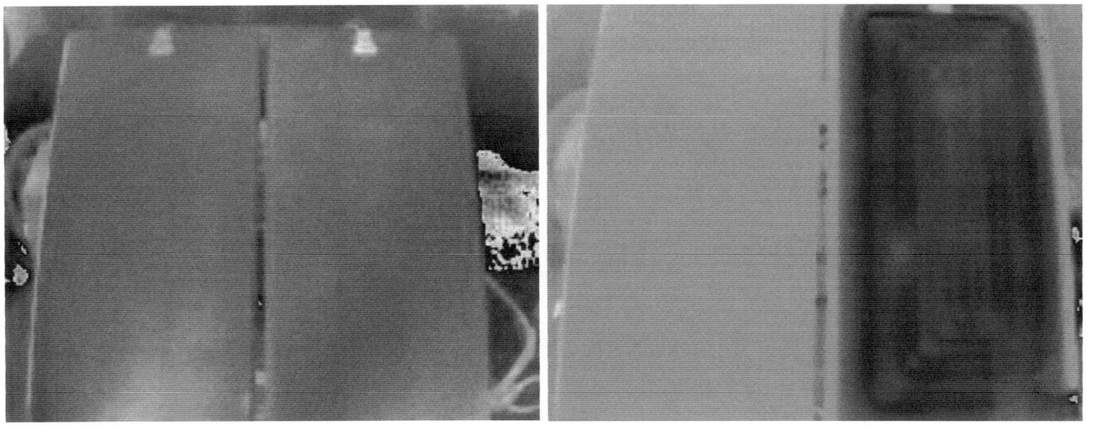

**Figure 7.** Temperatures of GFC elements without active TS layer (left) and with active TS layer (right).

Figures 6 and 7 show the UV diagnostics of developed roof elements. The pictures on the left shows roof elements with the deactivated HT layer and both modules were without the active PV layer. The pictures on the right show panels with a clearly visible active heat exchange layer, which has a significantly lower temperature due to flowing liquid (bright spiral on purple background).

An important aspect of quality control of active roof modules is the functionality of the PV layer. It is necessary to check whether there was any damage during production or installation, or in time due to external climatic influences. Such damage can be detected by a diagnostic electroluminescence method.

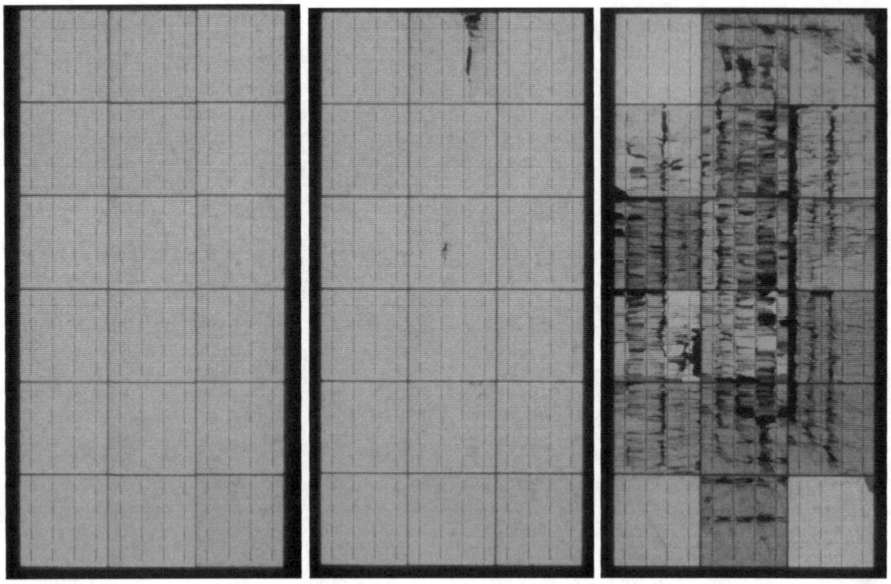

**Figure 8.** Damage to the PV module displayed by electroluminescence method, undamaged PV module (left), damaged PV module (right).

Figure 8 shows the possible damage to PV modules displayed by the electroluminescence method. Such damage can occur during winter season when water is used as the heat transfer medium in the HT layer. Cracks are visible in the figure and the pipeline that deformed the PV layer due to frost is clearly visible. Despite this considerable damage, the PV layer generated power only about 25% lower than before damage.

The actual methodology of energy efficiency evaluation is based on the conclusion of the applied comparative measurement method, where we used the possibility of measuring the parameters of PV modules and their energy outputs without the need to use information about spectrum and intensity of illumination The principle of the method is to compare the achieved electrical powers of the measured PV layers of GFC panels with the active and inactive HT layers.

### *3.3. Calculation of obtained heat from active HT layer*

In the actual measurement, periodically obtained heat from the active heat exchange layer was monitored in relation to the flow rate of the heat transfer medium through the heat exchange layer. The difference between input and output temperatures was also monitored. These data were entered into the calorimetric equation and the obtained heat was then calculated.

The calculation of the obtained heat from the activated heat exchange layer of the GFC panel with the PV layer is based on the Calorimetric equation $Q = m \times c \times (\vartheta 2 - \vartheta 1)$, where:

$Q$ = obtained heat [J],

$m$ = weight [$kg \times m^{-3}$],

$c$ = specific heat of water [J × K-1 × kg-1 ]

$\vartheta 2$= liquid output temperature [°C],

$\vartheta 1$= liquid input temperature [°C].

Then, from the above relationship, we apply the measured units to the simplified relationship. The input temperature of the GFC panel $\vartheta 1$ [°C], the output temperature of the GFC panel $\vartheta 2$ [°C], the average flow rate V [liters × $sec^{-1}$] and the specific gravity of water m = 991 [$kg \times m^{-3}$] are measured. The heat obtained is calculated as follows:

$Q = V\times(991\times10\text{-}3)\times4{,}185\times(\vartheta 2 - \vartheta 1) =>$

$Q = V\times4{,}147.3\times(\vartheta 2 - \vartheta 1)$ [J]

### *3.4. Static assessment of the system, measurement of wind resistance*

The described stage summarizes the most important findings, results of measurement and evaluation of resistance tests, ie. the load-bearing capacity and serviceability of the developed roof elements under wind loads. The assessment of PV roof elements in terms of resistance to wind was carried out in cooperation with VUT FAST university.

The subject of the load tests were concrete roof panels with a photovoltaic layer with dimensions a x b = 1,050 x 845 mm and the thickness of the load-bearing, ie concrete part of the panel structure (without photovoltaic layer) d = 20 mm. The aim of the load tests was to verify the load-bearing capacity of the panels in order to determine the recommendations for the ultimate load from the effects of suction or wind pressure based on the test results.

In case of loading of the roof element by wind suction, the safe recommended design value of the surface uniform load was equal to pd = 3.75 kN/m$^2$, based on the performed tests. In the case of a roof element loaded with effects simulating wind pressure, the safe recommended design value of the surface uniform load was equal to pd = 3.80 kN/m$^2$. Performing of the test is shown in figure 9.

**Figure 9.** Wind load test of developed roof element.

A complete static calculation was also performed during this stage. The assessment was performed according to ČSN EN 1991-1-3, for the roofs of buildings for the Czech Republic snow areas, class I to VII. The results showed that the construction of the developed roof element is suitable for I. and II. limit state in terms of static assessment.

## 4. Conclusions

The effort of project solvers is to achieve entirely new economic efficient using solar energy by means of energy active roof covering. Current research and tests proved, that there is a possibility to create such a system using present materials and technological solutions

Numerous laboratory testing tasks were accomplished at the beginning of the project to design and verify suitability of chosen materials for active roof cladding system production. During the process, the technological solution of individual tile structure was developed, including profiled GFC element with integrated HT layer bearing a system of PV cells, joint together by carefully chosen silicon putty, which effectively joints and seals the whole structure. All the components needed for manufacturing the roof tiles were purchased. In laboratory conditions, different GFC recipes were tested to find the most suitable composition.

Research works in 2017, which are not mentioned in this paper, included testing of the individual parts of the roof system, their connection and composition, mostly from a long-term point of view. Testing criteria were technological, physical-mechanical, hydrophobic and visual. Strength characteristics of the profiled GFC elements, adhesion of the PV layer to the GFC element, water-resistance and climate influence (temperature fluctuation, UV radiation, etc.) of the whole roof cladding system were tested. Electric parameters and temperature coefficient were measured.

In following years 2018–2019, technical realization of the whole system was made, where we mainly focused on achieving the highest functionality – durability, effectivity and energetic efficiency. Static and fire-resistance evaluation of the system was also conducted. Interconnection of individual tiles was designed and electric and heat-transfer efficiency was studied and calculated. Some examples are shown in the text above.

The authors' efforts aim was to develop a completely new system of economically effective use of solar energy. Electric energy acquired will be used immediately, while acquired thermal energy can be stored for future use. The developed system can be installed on both civil and commercial buildings. The results achieved during this project will be protected by copyright and commercialized both in domestic market and abroad until 2022.

**Acknowledgments**
The authors wish to express their gratitude and sincere appreciation to the authority of The Ministry of Industry and Trade of the Czech Republic, project No. FV10297 for financial support.

**References**

[1] Leber P, Vlach L, Bibora P and Čechmánek R 2016 Active Roof Cladding System Reducing Energy Consumption of Buildings *Annual report of the project No. FV10297* (Brno)

[2] Drdlova M, Cechmanek R and Ridky R 2015 Blast impact behaviour of concrete with different fibre reinforcement *11th International Conference on the Mechanical and Physical Behaviour of Materials under Dynamic Loading (DYMAT) Lugano, Switzerland, EPJ Web of Conf.* Vol. **94** Art. No. 05006

[3] Cechmanek R, Drdlova M and Prachar V 2018 New perspective material for building envelopes, *21st International Conference on Building Materials, Products and Technologies (ICBMPT), Ceskovice, Czech Republic, IOP Conference Series-Materials Science and Engineering* Vol. **379** Art. No. 012039

# Mitigation possibilities of concrete pavement degradation

**J Grosek[1], T Zavrel[1] and J Stryk[1]**

[1] Department of Transport Infrastructure, Transport Research Centre – CDV, Lisenska 2657/33A, 63600 Brno, Czech Republic

E-mail: jiri.grosek@cdv.cz

**Abstract**. Monolithic as well as precast Portland Cement Concrete pavement have shown a gradual increase in quantity and severity of defects and failures in recent years. Especially, the radical increase in the auto-destructive deterioration of the concrete structure, which is most often described as the expansion reaction, is alarming. The consequences may lead to extreme reduction of the lifetime of concrete on loaded or unloaded surface courses. The main causes are due to the combined effect of environmental effects (climate, chemical pollution), drainage of roads and the behaviour of the existing cement binders, provided that we do not reflect the modification of pavement structure layers. Only the impact of climate effects can be effectively reduced on the existing roads. The paper presents a summary of results obtained from the current portfolio of secondary concrete protection products. The testing was performed in terms of resistance to water infiltration and durability against the common effects of freeze/thaw and salt. The results show a dependence or independence according to the material basis when exposed to secondary protection products. Oils, silicate emulsions and a group with other material bases, so called other products, behave quite differently.

## 1. Introduction

A cement concrete pavement is a directly overrun construction layer within road pavement layers. However, surfaces on old concrete as well as on young, not resistant, concrete are at risk to fast local or full-surface material degradation [1, 2]. The extension of durability on construction sites is possible with full-surface protection. The surface on concrete pavements must be treated in order to keep their behaviour on roads [3, 4]. There are common influences caused by traffic loading or environmental erosion, physical and chemical reactions, such as an impact of chloride ions and freeze-thaw cycles, alkali-silica reactions, etc. [5]. The resistance of these deteriorations depends on concrete pavement quality, technological imperfections and conditions on site. Concrete surfaces are directly exposed to climatic conditions and have a contact with potentially aggressive substances. Concrete is a material that is hydrated for a long time for dozens of years, even though this process is reduced over time. Therefore, it needs a source of free water for its ongoing hydration. The concrete structure allows penetration or drying and movement of moisture (water). The movement is unstable over time and penetrates the system of pores, cracks and capillaries based on the current vectors of moisture and thermal gradient [6, 7].

Moisture in concrete has an effect on the vaporization and watering of the overrun surface, on drainage at the joint with subbase and moisture level at the joint with subbase at the interface with the road subbase, and furthermore, on drainage of the whole road pavement at the subgrade level. The moisture level is also significantly affected by temperature. With decreasing temperature there is a risk of condensation from air humidity, and at freezing temperature there is a risk of a part of free water in the structure getting frozen. The part affected by frost changes with the current position of the dew point temperature in the road structure [8, 9].

Water, or moisture, is in a certain degree necessary for concrete. However, fast and significant changes in moisture in concrete and long-term surplus or long-term deficiency of moisture (water) is harmful. Significant changes in the amount and vectors affecting moisture gradients may lead to hydration product wash-out of concrete or, in contrast, to introduction of substances from the environment into concrete. In both cases the ongoing hydration could be affected.

Potential solutions regard secondary protection means. Some of them effectively reduce the magnitudes of temperature gradients by reducing water permeability while maintaining the steam permeability of concrete surfaces. Unless the amount of penetrating aggressive substances can be sufficiently reduced in time, they become reactive and harmful in the concrete structure. In those cases, we need to maintain the concrete structure in dry condition, if possible. This is not always easy, since the water source may not only come from the surface, but also from joints and other elements, or it may be in the form of capillary moisture from the bottom layers of the road pavement structure. In those cases, it is individual for each point in structure when the secondary protection means are still so effective that they could minimize expansive reactions in the concrete structure through moisture reduction, and when they are not. The requirements for the minimization of moisture changes, together with the requirement for resistance to aggressive substances, as well as requirements for resistance to frost and chemical defrosting agents, are two crucial, usually completely different, requirements that should be the technological key for selecting a specific concrete secondary protection means.

## 2. Materials and specimens

Concrete mixture and cube specimens (100 mm molds) were prepared in the laboratory. The mix composition used in this testing was prepared in accordance with the Road and Motorway Directorate of the Czech Republic mixing design for the exposed aggregate construction of highway surface. The selected water-to-cement ratio was 0.45, and the utilized cement was CEM I 42.5 R, which was manufactured by Heidelberg Cement Group. The coarse aggregate was limestone, gradation category Gc 90/15, with nominal maximum aggregate size of 8 mm. Fine aggregate was river sand conforming to ČSN EN 933-1 and ČSN 736123-1 specifications [10, 11]. No concrete pavement admixtures (e.g. air-entrained, superplasticizer additive, etc.) or other chemical substances were used to distort the results of the water absorption and coating effectiveness (table 1).

Ten different products of coating were selected. They were marked with numbers 1 to 10, from the available offer. Number 10 is not evaluated for its extreme viscosity, which prevents coating by sprayers. The products were applied manually by brush by one thin layer (figure 1).
Description of products:

- oil based products, numbers 4 and 9,
- silicate based products, numbers 2, 5, 6, 7, 8,
- products with other material bases, numbers 1, 3.

The secondary protection products were provided by local suppliers in Central Europe. Unfortunately, it is not possible to specify the exact product compositions, or it is not published here on request of suppliers. The material basis was determined only by the macroscopic description and classified into the groups mentioned above.

**Table 1.** Mix composition of specimens.

| Component | Density (kg/m$^3$) | Locality |
|---|---|---|
| Cement | 420 | Mokra |
| Aggregate 0/2 | 692 | Tovacov |
| Aggregate 4/8 | 1138 | Nemojov |
| Water | 189 | - |
| w/c ratio | 0.45 | - |

The results are subsequently converted and put into a graph as a relative value of water saturation (infiltration) against the same and equally tested specimen without any secondary protection "REF Concrete" (see paragraph 4.1).

**Figure 1.** Cube specimens coated with products of secondary protection:
a) oil-based products, b) other based products, c) silicate-based products.

## 3. Experimental methods

The currently effective standard for concrete is ČSN EN 206+A1 [12]. This standard works with a term "working life" and "environmental actions". Regarding expansive reactions, the alkali-silica reaction of aggregates and reactiveness of carbonate aggregates are mentioned. Concrete resistance to effects of water, frost and defrosting agents is ensured by a right selection of environmental actions; in the Czech Republic the testing is performed according to ČSN 73 1326 (potentially according to ČSN 73 1322 and, if required, according to EN 480-11). Secondary protection is not a separate article according to this standard [13–16].

Regarding secondary protection, the closest to concretes is EN 1504-2 "Products and systems for the protection and repair" [17]. It uses terms "Hydrophobic impregnation", "Impregnation" and "Coating". The standard concerns all concretes and mentions five suitable principles. For the needs of cement concrete pavements, the crucial principles are principle 1 (PI – protection against infiltration) and principle 2 (MC – moisture control). Less important are usually principles 5 (PR – physical resistance), 6 (RC – resistance to chemicals) and 8 (IR – increasing resistivity).

The following tests are related to recommended tests for principles 1 and 2:

- water infiltration rate,
- water steam permeability,
- water penetration depth.

In order to compare the effect of secondary protection against water penetration into concrete, we designed our own model test "Water infiltration rate in concrete". The test is performed on laboratory specimens that are, at the age of 28 days, dried to the minimum weight, weighed and then left for at least 7 days under water up to the maximum weight, and weighed again. After keeping them at 20°C and relative humidity of 50%, "balanced" moisture is reached, and coating is performed along the whole perimeter by a secondary means. After the coating dries, the specimen is ready for the test itself, which is based on the immersion for 2 hours in water, wherein it is continuously weighed every 15 minutes by laboratory calibrated scale with an accuracy of 0.01 g (figure 2a).

Afterwards, the test "Water infiltration rate in concrete" on the same specimens was accompanied with tests in the environment of frost and salts according to the existing ČSN 73 1326 in KD-20 chamber, i.e. with specimens pre-saturated in water (figure 2b).

**Figure 2a.** Water infiltration rate testing of cube specimens coated by products of secondary protection (specimen coated by silicate-based product No. 2).

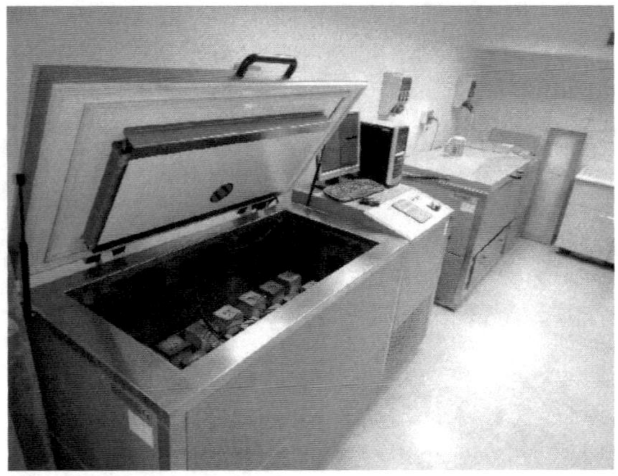

**Figure 2b.** Resistance of cement concrete surface to water and defrosting chemicals testing, KD-20 chamber with all specimens.

## 4. Results and discussion

### *4.1. Water infiltration*

In literature, water infiltration is accepted as a parameter related to concrete durability. The water absorption by immersion gives an estimate of the total pore volume of the concrete surface, but gives no indication on the concrete permeability, which is more important regarding durability [3, 18]. This testing method can be used to verify short-term surface water absorption and coating agent effectiveness, since water and other applied agents may cause the concrete destruction [5]. If we ensure lower absorption of concrete, we can increase its resistance. Absorption tests could be used as a quick testing method when the results provide us with an idea of the total water infiltration content over a certain time period [18].

The result of "Water infiltration" test is shown in figure 3. Figure 3 shows the percentage of relative water saturation differences between values of each individual product during the test after 15 minutes. Water saturation was measured using precision laboratory calibrated scale for two hours. Results showed that all coating products reduced water saturation of the concrete surface but with different effectiveness in comparison to the reference specimen of ordinary Portland Cement Concrete (REF).

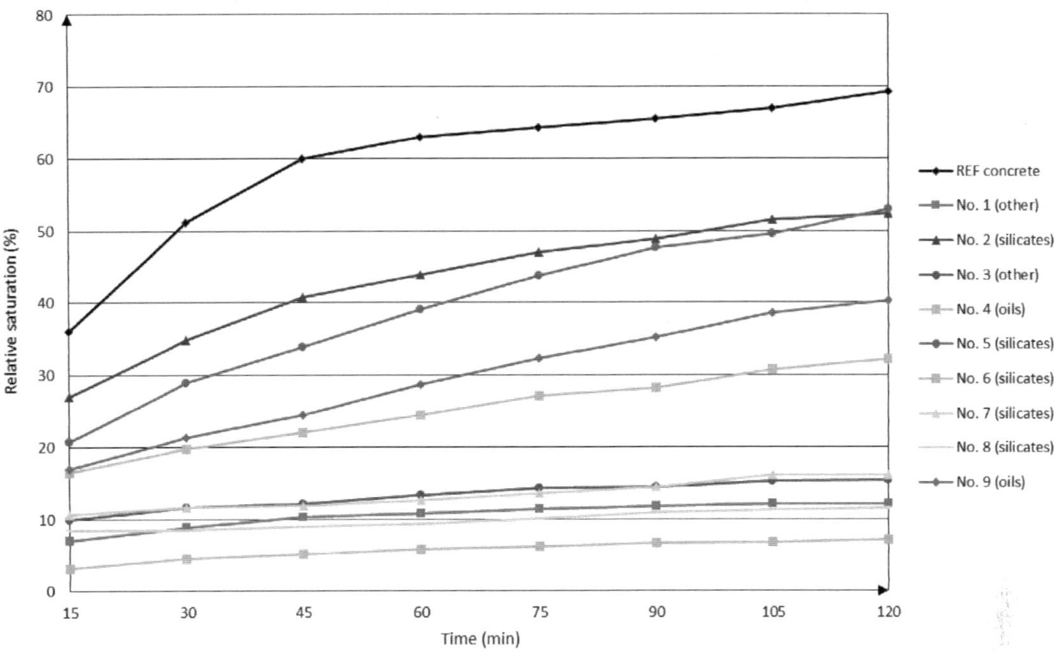


**Figure 3.** Water infiltration rate results for all groups: silicates (No. 2, 5, 6, 7 and 8), oils (No. 4 and 9) and other material bases (No. 1 and 3).

Generally, we cannot say that some material bases have equal or at least similar water intrusion characteristics as well. The "silicates" groups (No. 2, 5, 6, 7 and 8) have extremely different effectiveness, which is probably due to the composition of each product. On the other hand, it appears that the "oil" products (No. 4 and 9) have similar effectiveness.

### 4.2. *Freeze-thaw resistance*

It is expected that penetration of impregnated matrix by destructive agents (e.g. chloride ion) may also be reduced in comparison with the unimpregnated matrix. In contrast, several studies have showed that effectiveness of different types of surface treatment materials may vary depending on the exposure conditions. For example, a study by Suleiman, et al. showed poor performance under sulfate attack on concrete protected by a silicate agent. Another study showed that using silane as a surface treatment material enhanced the durability of concrete that was fully immersed in a sulfate solution compared to sodium silicate [3]. Therefore, caution is required to select the appropriate concrete coating material for different exposure conditions. Regarding concrete pavement, freeze-thaw resistance against defrosting chemicals is important.

The results of freeze-thaw tests show that the resistance of coated specimens is higher than the non-coated ones. Coated specimens are generally considered more effective based on the reduction of the amount of penetrated chloride ion, therefore, the destruction of concrete after testing is not too fast and extensive [1, 18].

The total mass of scaled material related to the test surface after each measuring interval is evaluated, in g/m$^2$. Measurements of surface scaling are performed at the beginning of the freeze-thaw test (0 cycle) after 25$^{th}$ and 50$^{th}$ freeze-thaw cycle. The scaled material from the specimen surface after 25(50) freeze-thaw cycles in 3% NaCl solution is shown in Figures 4a and 4b. According to freeze-thaw test results, the lowest surface scaling was obtained from oil applied specimens (No. 4 and No. 9). We can say that the "silicates" groups (No. 2, 5, 6, 7 and 8) have extremely different resistance results.

Figure 4a. Scaling of concrete surface after 50 freeze-thaw cycles (specimen coated by silicate-based product No. 2).

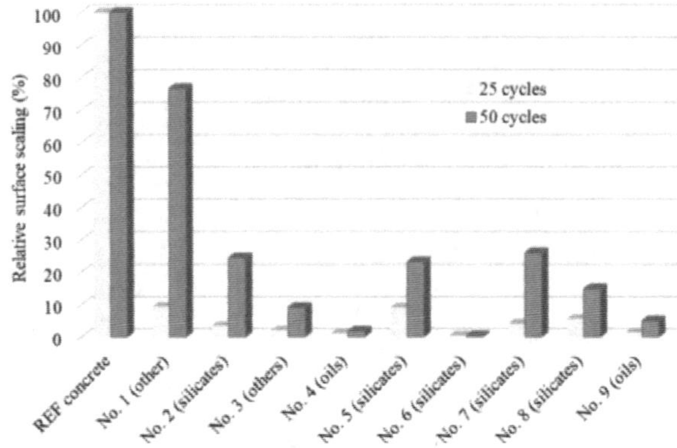


**Figure 4b.** Relative scaling of concrete surface for all groups: silicates (No. 2, 5, 6, 7 and 8), oils (No. 4 and 9) and other material bases (No. 1 and 3).

## 5. Conclusions

There is a huge number of different types of impregnation materials and techniques which should be investigated in further studies in relation to finding smart solutions for pavement maintenance. This article deals with the influence of secondary protection products of concrete pavement and different effectiveness of selected parameters. Researched products were composed of oils, silicate emulsions or other "non-detected" bases and the impact on water absorption and freeze-thaw resistance of laboratory specimens was monitored. Based on the results, it can be concluded that:

- According to laboratory testing, it was found that there is no general dependence between the parameters of the concrete absorption and the resistance of cement concrete surface to water and defrosting chemicals. The currently offered portfolio of coating products is able to improve only a single monitored parameter but hardly both of them.
- Current testing has an insufficiently accurate test portfolio to verify secondary protections.
- Road diagnostics to determine the cause of failure must be performed individually for each pavement structure or construction to reach a decision which parameter we need to improve by the coating.
- Moreover, concrete surfaces of different ages may react differently to coating effectiveness.
- Secondary protection on the basis of linseed varnish oil proved to be very successful in the Czech Republic in the past but is not environmentally friendly. Therefore, new development of oil products has potential for research in the future. The long-term testing and monitoring of surface properties for traffic safety, e.g. anti-skid testing, needs to be used on highways.

## Acknowledgements

This article was produced with the financial support of the Ministry of Education, Youth and Sports within the programme of long-term conceptual development of research institutions on the research infrastructure acquired from the Operation Programme Research and Development for Innovations (CZ.1.05/2.1.00/03.0064).

## References

[1] Han Y M, Dong G S and Doo S Ch 2007 *Constr. and Build. Mat.* **21** pp 362–369
[2] Zhang M H and Gjorv O E 1991 *Cem. and Concr. Res.* **21(6)** pp 1006–1014
[3] Suleiman A R, Soliman A M and Nehdi M L 2014 *Constr. and Build. Mat.* **73** pp 674–681
[4] Moon H Y, Kim H S and Choi D S 2006 Relationship between average pore diameter and chloride

diffusivity in various concretes *Construction and Building Materials* **20(9)** pp 725–732
[5] Frybort A, Vsiansky D, Stulirova J, Stryk J and Gregerova M 2018 *Mat. Charact.* **137** pp 91–108
[6] Hodakova D, Zuzulova A, Capayova S and Schlosser T *SGEM 2017: Inter. Multidisc. Scien. GeoConf.* **17** pp 467–472
[7] Hodakova D, Zuzulova A, Capayova S, Schlosser T and Branis M 2018 *SGEM 2018: Inter. Multidisc. Scien. GeoConf.* **18** pp 419–426
[8] Chatterji S 1982 Probable mechanisms of crack formation at earlyages of concrete: a literature survey *Cem. and Concr. Res.* **15** pp 371–376
[9] Kesai Y, Matsui I and Yokohama K 1982 Shrinkage and cracking of concrete at early ages *Proc. of the Intern. Conf. on Concr. at Early Age* pp 45–50
[10] ČSN EN 933-1 Tests for geometrical properties of aggregates – Part 1: Determination of particle size distribution – Sieving method (Prague: UNMZ)
[11] ČSN 736123-1 Road building – concrete pavements – Part 1: Construction and conformity assessment (Prague: UNMZ)
[12] ČSN EN 206+A1 Concrete – Specification, performance, production and conformity (Prague: UNMZ)
[13] ČSN 73 1326 Resistance of cement concrete surface to water and defrosting chemicals (Prague: UNMZ)
[14] ČSN 73 1322 Determination of frost resistance of concrete (Prague: UNMZ)
[15] ČSN P 73 2404 Concrete – Specification, performance, production and conformity - Additional information (Prague: UNMZ)
[16] ČSN EN 480–11 Admixtures for concrete, mortar and grout – Test methods – Part 11: Determination of air void characteristics in hardened concrete (Prague: UNMZ)
[17] ČSN EN 1504–2 Products and systems for the protection and repair of concrete structures - Definitions, requirements, quality control and evaluation of conformity – Part 2: Surface protection systems for concrete (Prague: UNMZ)
[18] De Schutter G and Audenaert K 2004 *Mat. and Struct.* **37** pp 591–596

IOP Conf. Series: Materials Science and Engineering 1039 (2021) 012019 doi:10.1088/1757-899X/1039/1/012019

# Utilization of continuous hydrostatic weighing for monitoring of volume changes

**R Hermann[1], R Drochytka[1], V Cerny[1] and P Figala[1]**

[1] Brno University of Technology, Veveri 331/95,602 00 Brno, Czech Republic

Email: hermann.r@fce.vutbr.cz

**Abstract**. This paper deals with the problematics of utilization of continuous hydrostatic weighing to monitor the volume changes in research of new building materials. The aim of this paper is to describe the abilities of continuous hydrostatic weighing and to present the capabilities and results of measurements made using this method. The volume changes in this paper were measured using the laboratory scales connected to computer. The volume changes were studied on the cement-based composites containing up to 40 wt.% of cement substitution by secondary raw material. The high temperature fly ash, packing glass, foundry sand and slag were used as the cement substitution. This research shows that the continuous hydrostatic weighing produces results comparable to other methods for volume change measuring. Therefore, this method can be used for volume change monitoring without using any economically demanding laboratory equipment.

## 1. Introduction

The volume changes of cementitious composites, mortars, screeds and concrete are very known throughout the years. The cracking of cementitious mortars and concretes in the early stages of hydration are something, that concerns scientists as well as the public audience, because these sorts of failures are often visible by naked eye. Not at first of course, but the microcracks which are developed in the early stages of curing can become a bigger problem, because these cracks open and expose the structure of the matrix of building material. [1, 2] And therefore, they contribute to the reducing of the resistance of the material against exposure conditions. These microcracks which can occur in the very early stages of hydration can lead to structural problems and theoretically even to the collapse of construction. [3] The knowledge of the size of shrinkage or expansion of used building material is crucial for some applications. The volume changes can be reduced or increased by many additives and admixtures. The current aim of development is mainly to reduce the ecological and economical demands of the production of building materials. This can be achieved using secondary raw materials to substitute cement. Substitution of cement is effectively used in blended cements, which consists of clinker and other components. In last few years, many studies try to find use for waste materials and utilize them in building materials. Many products from power plants are being used for such purpose. This paper studies the effect of addition of some other types of secondary raw materials on the volume changes of cementitious mortar, for example furnace slag from power plant, foundry sand with soluble glass or finely ground waste glass. [4, 5]

## 2. Materials and Methods

In this experiment, the early volume changes of cementitious mortar were studied using the continuous hydrostatic weighing. Laboratory scales connected to the computer with application to log the data from scales were used.

### Raw materials

For this study, the following raw materials were used: Portland cement 42.5 R from Českomoravský cement, a.s. - Závod Radotín and various siliceous filler (sand) with the size distribution of 0.00–4.00 mm and ground limestone. The mix contained polypropylene fibers and admixtures (PCE plasticizer). As secondary raw materials the following materials were used. Furnace slag from non-functioning coal thermal power plant Oslavany (SSL), foundry sand containing soluble glass from ArcelorMittal Ostrava (SCSG), waste glass from Vetropack Moravia Glass (Glass) and high-temperature fly ash from coal power plant Tušimice (HFA).

The high-temperature fly ash and finely ground waste glass were used as a substitution of cement in rate of 20, 40 and 60 wt.%. The furnace slag and foundry sand were used as a substitution of filler in rate 100 wt.%.

### Pre-treatment of secondary raw materials

The secondary raw materials had to be pre-treated by crushing, milling and/or sieving. The furnace slag (SSL) was pre-treated by crushing and sieving to a fraction of 0.5–4.0 mm to substitute completely coarse aggregate in reference mixture. The foundry sand (SCSG) which was delivered in large hardened stones with diameter up to 30 cm had to be crushed and then sieved to separate the sand from iron. The SCSG was used to substitute the fine aggregate with fraction of 0.0–0.5 mm. The used waste glass was be pre-treated by milling to substitute primary binder, cement, which had specific surface of around 400 $m^2 \cdot kg^{-1}$ (± 50 $m^2 \cdot kg^{-1}$). HFA was also pre-treated by milling to achieve the corresponding specific surface of cement. For the pre-treatment of secondary raw materials, the jaw-crusher and laboratory ball grinder were used to achieve the corresponding fractions of aggregate and specific surfaces. In the table 1, the fraction distribution of secondary raw materials used for the substitution of cement (HFA and Glass) are shown.

**Table 1.** The fraction distribution of HFA and Glass after milling [wt.%].

| | <0.01 mm | *0.01-0.063 mm* | *>0.063 mm* |
|---|---|---|---|
| **HFA** | 28.84 | 63.35 | 7.87 |
| **Glass** | 30.71 | 53.12 | 16.18 |

### Mixing

Every mix-design containing cement in combination with SSL, SCSG, Glass, HFA, sand, limestone, PP fibres and dry PCE plasticizer was pre-mixed in laboratory four-speed mixer for 1 minute. Then the mixing water was added, and the mixture was then mixed for additional 3 minutes. At the end of each mixing period, the testing samples were cast, and two 1 kg fresh mortar samples were also poured into specialized PP bags which could be heat-sealed. These samples were then used for the measurement of autonomous volume changes by continuous hydrostatic weighing. The water/binder ratio was changed for each mixture correspondingly to achieve the same plasticity of fresh mortar (tested on flow table in accordance with CSN EN 1015-3). In the table 2, the mixing ratios are shown. [6, 7]

**Table 2.** Mixing proportions of raw materials samples [wt.%].

| | Cement | *Limestone* | *Sand 0.0-0.5 mm* | *Sand 0.5-4.0 mm* | *Glass* | HFA | SCSG 0.0-0.5 mm | *SSL 0.5-4.0 mm* | *w/b ratio* |
|---|---|---|---|---|---|---|---|---|---|
| **Reference** | 1.0 | 0.6 | 0.6 | 2.0 | - | - | - | - | 0.57 |
| **Glass** | 0.6 | 0.6 | 0.6 | 2.0 | 0.4 | - | - | - | 0.59 |
| **HFA** | 0.6 | 0.6 | 0.6 | 2.0 | - | 0.4 | - | - | 0.585 |
| **SCSG** | 1.0 | 0.6 | - | 2.0 | - | - | 0.6 | - | 0.57 |
| **SSL** | 1.0 | 0.6 | 0.6 | - | - | - | - | 2.0 | 0.62 |
| **SCSG, SSL, HFA** | 0.6 | 0.6 | - | - | - | 0.4 | 0.6 | 2.0 | 0.63 |
| **SCSG, HFA** | 0.6 | 0.6 | - | 2.0 | - | 0.4 | 0.6 | - | 0.58 |
| **SCSG, Glass** | 0.6 | 0.6 | - | 2.0 | 0.4 | - | 0.6 | - | 0.59 |

As shown in the table 2, the Glass and HFA were used to substitute the binder, in this case cement by 40 wt.%. The SCSG and SSL were pre-treated to fully substitute either coarse or fine aggregate in the reference mixture. For the purposes of study, several combinations of binder and filler substitutions were tested.

**Volume changes test procedure**

The volume changes of test samples were assessed by the continuous hydrostatic weighing. The principle is shown in the figure 1 [9].

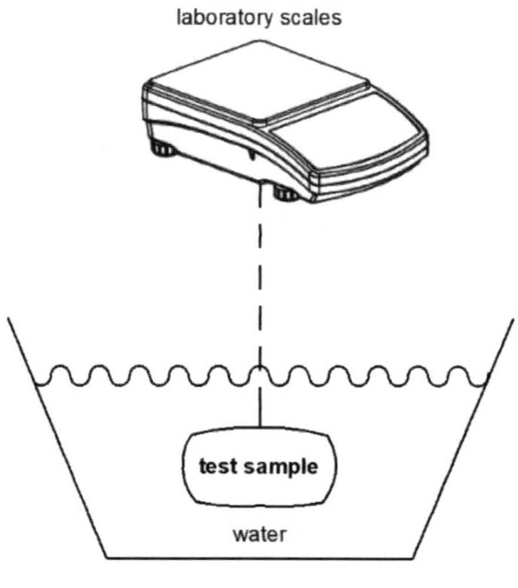


**Figure 1.** Illustration of hydrostatic weighing of test samples.

For the volume changes measurements, the RADWAG laboratory scales were used. These scales have automatic calibration in every two hours, which had to be disabled by the manufacturer to enable continuous measurement for days. Under these scales, large tank with water was prepared. Before the measurement, each test sample was weighed dry and the weight was registered. Each test sample was then attached to the bottom of the laboratory scales by a thin string and submerged in water to begin the measurement.

Each test sample was created by the usage of specialized kitchen type of bags for storing food, while vacuuming the excess air and heat-sealing the bag. After mixing of mortar, around 1 kg of fresh mixture was poured into the bag, vacuumed and heat sealed on two, to three places to ensure proper seal. Immediately after this procedure the bag was dry weighed and then hanged below the scales and submerged in water. Every test sample was then tested for 32 hours. The measurement data were transferred from scales into computer program and then evaluated and processed to show the corresponding volume changes.

Thanks to the vacuuming of the air bubbles trapped inside and around the mixture, ideally all the volume changes that occur in the material, are the volume changes caused by chemical reactions and hydration. And, since the bag is sealed, the samples cannot dry out.

## 3. Results and discussion

The shrinkage of cementitious composites during the early stages of hydration is caused by the chemical processes which are the cause of hardening of concrete/mortar etc. In this study, eight series of measurements were performed to demonstrate how the proposed technique using the laboratory scales connected to a computer can be used in assessing the value of volume change in the early stages of hydration. This method can be used even for longer measurements of the autogenous shrinkage, since the test sample is trapped inside a closed PP bag, which does not let any moisture in or out and therefore the test sample does not change its weight, only its volume.

In the figure 2 the raw measurement data of volume changes of the test samples are shown.

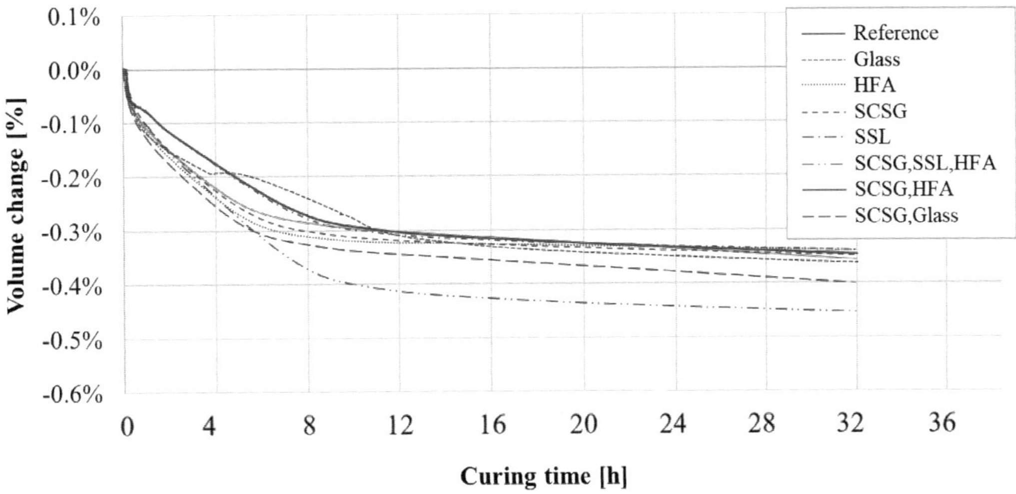


**Figure 2.** Illustration of hydrostatic weighing of test sample.

The volume changes were measured by the hydrostatic weighing and compared with volume changes measured by other methods to assess its usability in research. For every mix-design, 3 samples were used. The reference test samples were tested using the above method of continuous hydrostatic weighing and compared to test results using the shrinkage cone method according to the ASTM C827/C827M-16, with the exception, that the mixture in the cone was covered by plastic sheet to prevent drying of the test sample.

The volume changes measured by the hydrostatic weighing are represented as shrinkage percent. The value of shrinkage, for example, of the HFA test sample in the figure 2 was -0.34% after the 32 hours of curing. Since the percentage can be also written as the length change, this value can be also expressed as -3.4 mm/m. The volume changes of all test samples were recorded for 32 hours since mixing.

All recorded volume changes are summarized in the table 3, presented as the mm/m values to correspond with the results from other researches.

**Table 3.** Volume changes of tested samples [mm/m].

| | Volume change [mm/m] |
|---|---|
| **Reference** | -3.6 |
| **Glass** | -3.9 |
| **HFA** | -3.4 |
| **SCSG** | -3.6 |
| **SSL** | -4.0 |
| **SCSG, SSL, HFA** | -4.5 |
| **SCSG, HFA** | -3.6 |
| **SCSG, Glass** | -4.0 |

As shown in the table 3, the test results show, that the addition of any secondary raw material, both as the substitution of binder or filler lead to increase in shrinkage in the first 32 hours of curing, except for the HFA test sample containing 40 wt.% substitution of cement by high temperature fly ash. This corresponds with the results of Guangcheng et. al. (2006). In their research the effect of substitution of cement by silica fume, pulverized fly ash and furnace slag was tested. The results show, the substitution of cement by pulverized ash lead to a small increase in shrinkage, almost comparable to the reference, but the substitution by furnace slag lead to greater shrinkage. [5]

In the figure 3 the relation between the w/b ratio and the shrinkage of the tested samples.

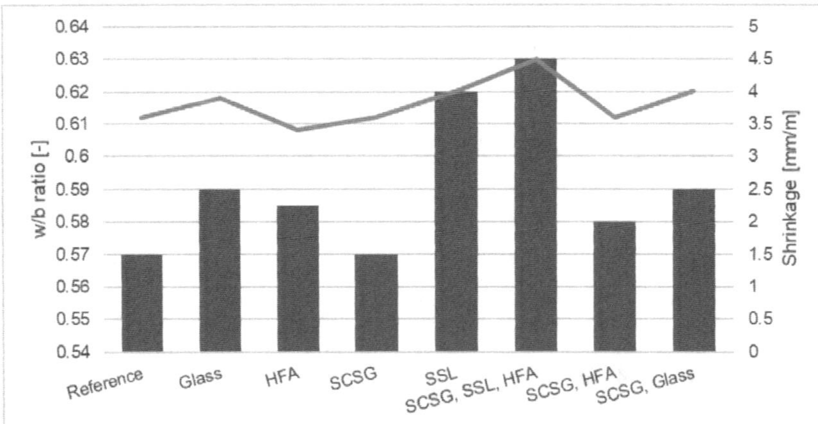


**Figure 3.** Illustration of the relation between the w/b ratio (blue columns) and the shrinkage (orange line).

In the figure 3, it is shown the w/b ratio affects the value of autogenous shrinkage of the tested samples. However, the relation is not linear for every type of tested secondary raw material and waste. The relation is not so direct when the HFA and SCSG was used.

## 4. Conclusion

Results presented in this paper show, that the proposed technique of measuring of volume changes by the continuous hydrostatic weighing is suitable for cementitious materials and is fully capable of measuring of the autogenous volume changes when the mixture is vacuumed and heat-sealed in PP bags. The presented technique does not require any highly sophisticated measuring devices and the measurements can be done by using only the standard laboratory equipment. The laboratory scales with resolution of at least two decimal places (0.00 gram) and ability to connect and transfer the measurement data to computer.

In this paper, the effect of substitution of binder and/or filler in cementitious mortar by the secondary raw materials and wastes on the volume changes in the early phases of curing was tested. The results show that any substitution of binder or filler in the mixture by the tested secondary raw materials and wastes lead to the increase of shrinkage in the early stages of curing. Only the substitution by the high temperature fly ash can lead to lower shrinkage in the early stages of curing. The results also show that there is some relation between the w/b ratio and the amount of shrinkage. The relation is linear when using the SSL and Glass, but nonlinear when using the SCSG or HFA.

The above problematics will be investigated in further research.

**Acknowledgement**
The paper was elaborated as a part of solution of project nr. FAST-J-20-6423 "Study of physical-mechanical properties of new siliceous and polymer composites with focus on volume changes monitoring".

**References**

[1] Aïtcin P C 2008 *Binders for durable and sustainable concrete* (New York: Taylor & Francis) ISBN 978–0–203–94048–8
[2] Aïtcin P C and Mindess S 2011 *Sustainability of concrete* (New York: Spon Press) ISBN 978-0-203-85663-5
[3] Choo B S and Newman J 2003 *Advanced concrete technology* (Amsterdam: Elsevier) ISBN 07-506-5104-0.
[4] Litoš J and Hošek J 2005 *Microstructure changes and volume changes of concrete components in a setting phase Cement Combinations for Durable Concrete* (Thomas Telford Publishing) pp 391–398 ISBN 0-7277-4004-0
[5] Guangcheng L, Youjun X and Zhengwu J 2006 Volume changes of very-high-performance cement-based composites *Magazine of Concrete Research* vol **58**(10) pp 657–663
[6] Radocea A 1998 Autogenous volume change of concrete at very early age *Magazine of Concrete Research* vol **50**(2) pp 107–113 ISSN 0024-9831
[7] ČSN EN 1015-3 Methods of test for mortar for masonry - Part 3: Determination of consistence of fresh mortar (by flow table) 2000 (Prague)
[8] ASTM C827/C827M-16 Standard Test Method for Change in Height at Early Ages of Cylindrical Specimens of Cementitious Mixtures 2016 (West Conshohocken: ASTM International, 100 Barr Harbor Drive, PO Box C700, West Conshohocken, PA 19428-2959, United States)
[9] ČSN EN 1015-10 Methods of test for mortar for masonry - Part 10: Determination of dry bulk density of hardened mortar 2000 (Prague)

IOP Conf. Series: Materials Science and Engineering 1039 (2021) 012020 doi:10.1088/1757-899X/1039/1/012020

# Impedance spectroscopy measurement of metakaolin-based alkali-activated building materials

**I Kusak[1,2], M Lunak[1], C Mizerova[2] and P Rovnanik[2]**

[1] Institute of Physics, Faculty of Civil Engineering, Brno University of Technology, Veveri 95, 602 00 Brno, Czech Republic
[2] Institute of Chemistry, Faculty of Civil Engineering, Brno University of Technology, Veveri 95, 602 00 Brno, Czech Republic

Email: kusak.i@fce.vutbr.cz

**Abstract.** Cement-containing as well as cement-free building materials are regarded as dielectrics. Therefore, electrically conducting admixtures are to be added to them in order to increase their electrical conductivity. Steel or carbon fibres, metal powder, graphite, carbon soot or carbon nanotubes are commonly used for this purpose. The conductivity increase offers new application options, such as sensor property materials, self-heated materials, or electromagnetic smog shielding materials. The specimens of the mixes to be studied were subjected to an electrical analysis carried out within the frequency range from 100 MHz to 3 GHz by means of an ZNC vector analyser and an SPEAG-made DAK-12 coaxial probe and, furthermore, a dedicated automatically measuring device operating within the frequency range from 40 Hz to 1 MHz. The frequency spectra of interest were measured on various copolymer specimens differing from each other by the content of graphite and carbon nanotubes. Higher content of these admixtures increases the electrical conductivity and the building materials thus become easier to measure by means of electromagnetic measuring methods.

## 1. Introduction

The development and the study of cement-containing as well as cement-free composites featuring increased electrical conductivity can be noticed, above all, during the recent two decades. The research focuses mainly on the behaviour of materials in electric fields, both DC and variable frequency AC ones, and also on the determination of the material properties which are closely related to the electrical conductivity change [1–3]. Water glass, NaOH and $Na_2CO_3$ is most frequently used as alkali activator. As far as their mechanical properties are concerned, the alkali-activated materials are comparable with the PC (Portland cement), particularly in the cases where the water glass is used as activator. These materials also feature better endurance, better resistance against aggressive environment, better frost resistance and lower amount of hydration heat released. Among their most important drawbacks there are the high autogenous as well as dry-up-related shrinkage, which is particularly characteristic for activators in the form of water-glass. If NaOH and $Na_2CO_3$ are used, their shrinkage is similar to that of PC [4].

Our results show that the application of carbon nanotubes in materials results in improving their strength, elasticity and a general durability [5] The nanotubes feature a low density (1.3-1.4 $g{\cdot}cm^{-3}$ depending on the nanotube type), a high thermal conductivity (1 750 to 5 800 $W{\cdot}m^{-1}\ K^{-1}$) and – thanks to delocalized bonds along the entire carbon layer – excellent electrical conductivity. Carbon nanotubes are considered to be a potential replacement of the composite reinforcement, their mechanical, electric and thermal properties exceeding those of the traditional fibers [5, 6]. The research results also show that high length-to-diameter-ratio fibrous fillers can change the concrete detachable capabilities at considerably lower concentrations than the particulate ones. However, attention is to be paid to the dispersion of fibrous fillers in comparison with the particulate ones, as the fibrous fillers are more difficult to disperse and easier to damage during the manufacture process [7].

## 2. Experiment setup

Following materials were used to prepare the graphite-admixture (MKG: metakaolin-graphite) test specimens: Mefisto K05 metakaolin, PMM 11 graphite powder, sodium water glass ($M_S = Na_2O/SiO_2 = 1.6$), PG1-3 normalized quartz sand, Triton X-100 ion-free detergent (2% solution), Lukosan S defoaming additive (1% solution) and water. First, PMM 11 graphite powder, Triton X-100 and a part of the water amount were added to the water glass. This mix was stirred in a mixer for three minutes. Subsequently, metakaolin (Mefisto K05), the remaining amount of water and 3 filler gradings (PG1-3 sand) were added. Finally, Lukosan S defoamer was added. From each of eight mixes with different graphite content, a set of three bodies of dimensions 40 × 40 × 160 mm each was made. Once the specimens were demoulded, they were wrapped in a foil and deposited in the laboratory for a period of 28 days. Subsequently, the specimens were dried up at a temperature of 105°C until their constant mass was reached. The measurement results are compared with those obtained on reference formula, graphite-powder-free specimens.

Following materials were used to prepare the test specimens with carbon nanotubes added (MKCNT: metakaolin-carbon nanotubes): Mefisto K05 metakaolin, carbon nanotubes (CNTs), Graphistrength CW2-45 (1% solution), sodium water glass ($M_S = Na_2O/SiO_2 = 1.6$), PG1-3 normalized quartz sand and water.

**Table 1.** Composition of graphite-added mixes (MKG).

| | REF | G2 | G4 | G5 | G6 | G8 | G10 | G12 | G15 |
|---|---|---|---|---|---|---|---|---|---|
| **Metakaolin (g)** | 350 | 350 | 350 | 350 | 350 | 350 | 350 | 350 | 350 |
| **Sodium silicate (g)** | 350 | 350 | 350 | 350 | 350 | 350 | 350 | 350 | 350 |
| **Graphite powder (g)** | - | 7 | 14 | 17.5 | 21 | 28 | 35 | 42 | 52.5 |
| **Sand (g)** | 1 050 | 1 050 | 1 050 | 1 050 | 1 050 | 1 050 | 1 050 | 1 050 | 1 050 |
| **Triton 2 % (ml)** | - | 7 | 14 | 17.5 | 21 | 28 | 35 | 42 | 52.5 |
| **Lukosan S 1 % (ml)** | - | 3 | 3 | 3 | 3 | 3 | 6 | 6 | 6 |
| **Water (g)** | 120 | 130 | 137 | 146 | 164 | 181 | 196 | 216 | 242 |

**Table 2.** Composition of carbon-nanotube added mixes (MKCNT).

| | REF | CNT0.05 | CNT0.10 | CNT0.15 | CNT0.20 |
|---|---|---|---|---|---|
| **Metakaolin (g)** | 350 | 350 | 350 | 350 | 350 |
| **Sodium silicate (g)** | 350 | 350 | 350 | 350 | 350 |
| **Sand (g)** | 1 050 | 1 050 | 1 050 | 1 050 | 1 050 |
| **CNTs solution 1 % (g)** | - | 17.5 | 35 | 52.5 | 70 |
| **Water (g)** | 120 | 102.5 | 85 | 67.5 | 50 |

First, a CNT suspension (1% CNT solution) and water were added to the water glass. This mix was stirred in a mixer for three minutes. Subsequently, metakaolin and 3 filler gradings (PG1-3 sand) were added to it. From each of the four mixes with different graphite content, a set of three bodies of dimensions 40 × 40 × 160 mm each was made. Once the specimens had been demoulded, they were wrapped in a foil and deposited in the laboratory for a period of 28 days. Subsequently, the specimens were dried up at a temperature of 105°C until their constant mass was reached. The measurement results are compared with those obtained from a reference formula specimen, which is CNT-free.

Three test bodies of dimensions 40 × 40 × 160 mm were manufactured for each of the conductivity admixtures. The results obtained from each test body are compared with those of the reference specimen.

The prepared samples were characterized by impedance spectroscopy in the range of 40 Hz to 1 MHz using an Agilent 33220A sinusoidal signal generator and an Agilent 54645A dual-channel oscilloscope. The output voltage of the signal generator was 5.5 V. The input values of the electrical capacitance and the resistance of the oscilloscope were 13 pF and 1 MΩ, respectively. These instruments were assembled for fully automated measurement. In order to perform impedance analysis, the prismatic specimens were placed between parallel brass plate electrodes (30 × 100 mm) so that the distance between the electrodes was 40 mm (figure 1.) [7].

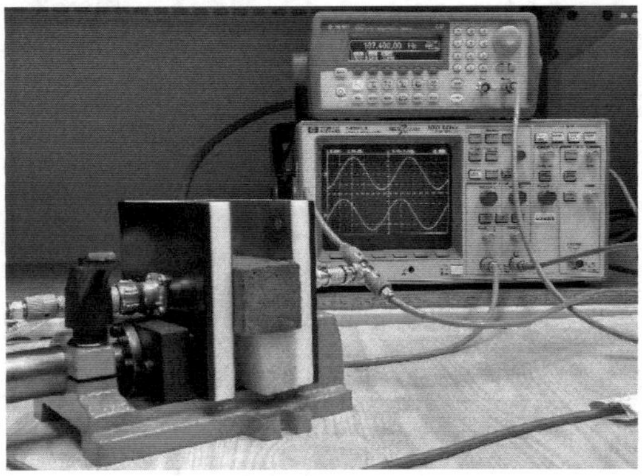

**Figure 1.** Experiment arrangement.

For higher range of frequencies, 100 MHz up to 3 GHz, R&S ZNC vector analyzer with DAK 12 coaxial probe manufactured by Speag was used. In this frequency spectrum the electrical conductivity and relative permittivity as a function of frequency have been measured.

## 3. Results

The Speag touch probe supplemented high-frequency analyzer has determined the specimen permittivity real components in the frequency range of 500 MHz to 3 GHz. For the reference specimens, the permittivity values begin at the value of about 2.3 at the lowest frequencies. When the frequency increases, they rise up to 2.8. Due to the CNT addition (figure 3), the permittivity levels grow up, whereas the curves for 0.05 and 0.1 additions intertwine with each other. The highest permittivity values, almost 3.5, were reached on specimens containing 0.2 CNT.

The specimens to which only graphite powder was added show elevated permittivity values, too (figure 2), reaching as much as 4.8. The lower permittivity values of carbon nanotube-added specimens are apparently due to the intertwined structure of the nanotube incorporation in the matrix, while the amount of isolated particles accountable for polarization effects is lower. The graphite powder containing specimens appear to contain more metakaolin-isolated particles, showing a polarization capability in the frequency range from 500 MHz to 3 GHz.

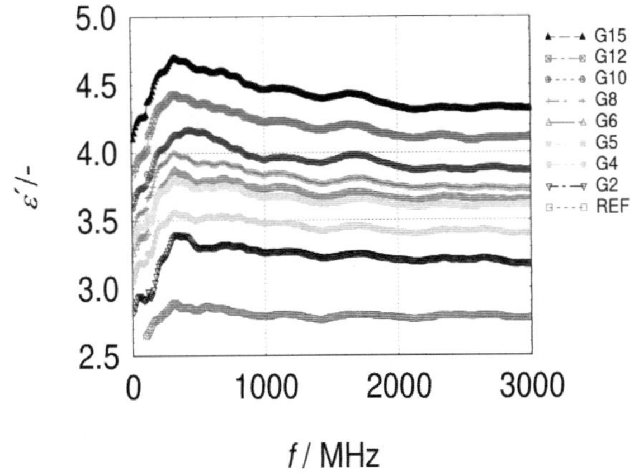


**Figure 2.** Permittivity real components for graphite-added specimens, as measured by means of Speag probe.

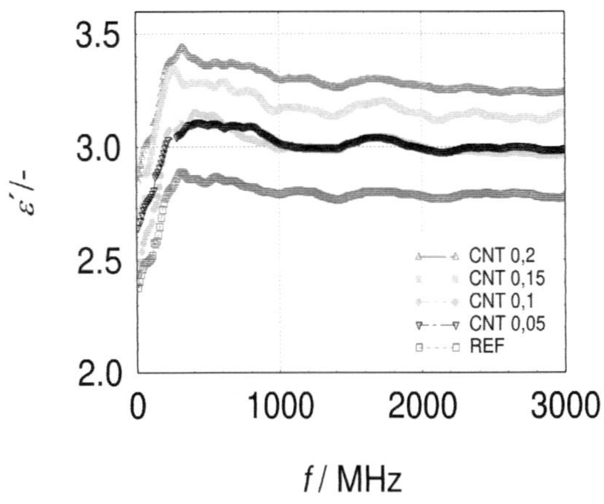


**Figure 3.** Permittivity real components for CNT-added specimens, as measured by means of Speag probe.

Different trends in the electric capacitance curves obtained by the impedance spectroscopy methods are observed in the lower frequency range from 40 Hz to 1 MHz. The capacitance of graphite-added specimens (figure 4) goes down, when the graphite amount grows up throughout almost the entire frequency range. The curves intertwine from G8 to G15, however, from 100 kHz upwards, the capacitance rises with the addition of G12, G15. The measurements results obtained in this frequency range (100 kHz to 1 MHz) are affected by a higher measurement error. Therefore, we do not attach relevance to them for the time being.

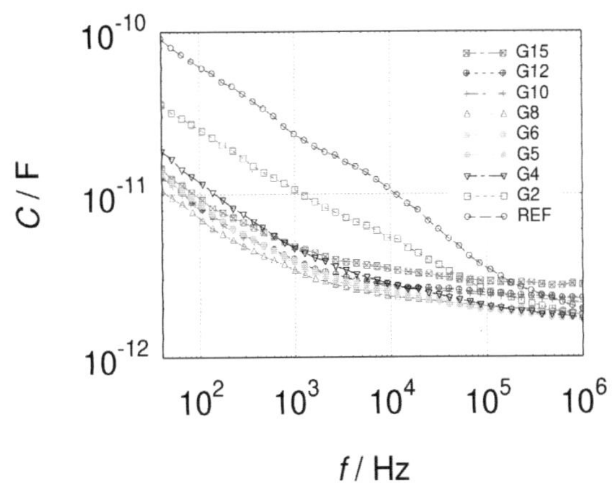


**Figure 4.** Capacitance of graphite-added specimens as measured by impedance spectroscopy method.

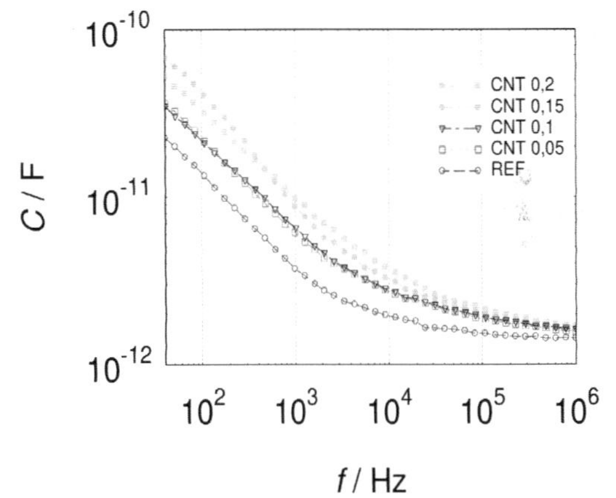


**Figure 5.** Capacitance of CNT-added specimens as measured by impedance spectroscopy method.

As expected for the carbon-nanotube-added specimens (figure 5), the capacitance grows up with increasing the amount of the nano-material in the specimen.

The unexpected capacitance fall shown by graphite-added specimens is explained by drying up the specimens containing metakaolin, which, as such, is predominantly composed of aluminium and silicon oxides. Due to the drying, conducting paths between carbon particles as well as possible conducting paths for ions in the matrix have disappeared. Absence of water thwarts the polarization in the low frequency range for the specimens used.

The electrical resistance of graphite-added specimens, as determined by means of the impedance spectroscopy method (figure 6), ranged from 100 MΩ at the lowest frequencies up to 10 kΩ in the 1 MHz region. The resistance spectra of graphite-added specimens grow up with the graphite content from 40 Hz to 10 kHz. At higher frequencies, the spectra intertwine, but the subsequent spectrum part is affected by a considerable measurement error. Prior to the measurements, the specimens were dried up. Together with the metakaolin, the graphite prevents the specimens from fast re-moistening. The specimens show a high DC resistance outwardly. For CNT specimens, the trend is reversed. Carbon nanotubes make the specimen electric resistance decrease (figure 7), the respective values being of the same order as those of the spectra of figure 6. The CNT strings create conducting paths in the metakaolin matrix, however, to a lesser extent. Compared to the resistance values for cement paste [7], the values of resistance were lower. It is not appropriate to compare the results of this article with [7], our recipes and results are new and original.

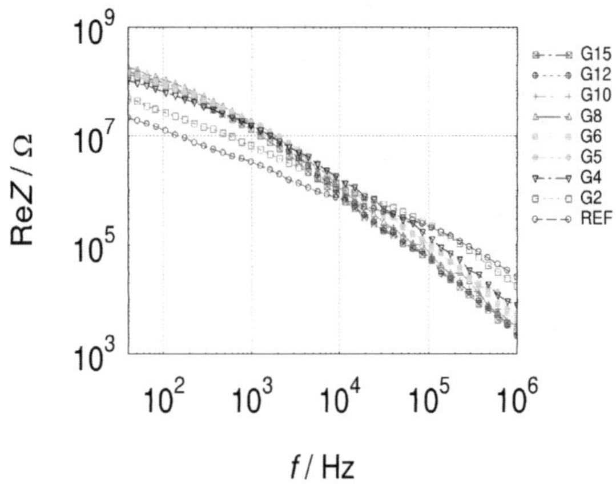


**Figure 6.** Impedance real component of graphite-added specimens as measured by impedance spectroscopy method.

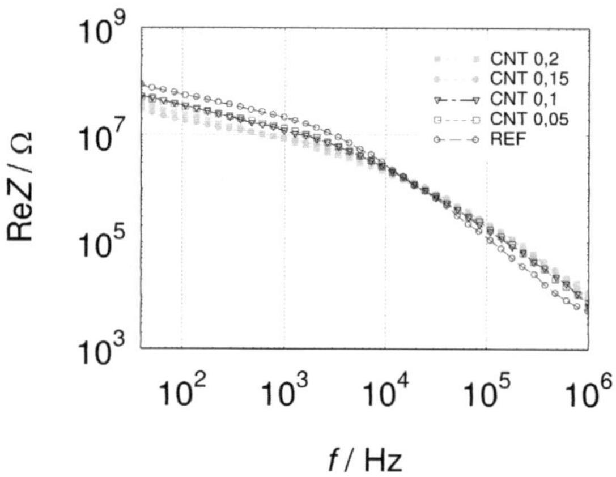


**Figure 7.** Impedance real component of CNT-added specimens as measured by impedance spectroscopy method.

## 4. Conclusion

The present paper deals with the change of electric parameters of metakaolin-based alkali-activated mixes. Graphite and carbon nanotubes were added to the different mixes. The electric conductivity of the materials in question is improved, which makes them easier to measure by electromagnetic-principle-based methods.

The metakaolin-based specimens have been measured by means of the impedance spectroscopy method in the low-frequency region and by means of a vector analyzer with a touch probe in the high frequency region. The graphite-added and CNT-added specimens featured a high electric resistance, corresponding to the specimen drying up and, in spite of several-day-storage in laboratory environment, a negligible wetting, which would improve the measurability of electrical quantities.

Pronounced changes in the spectra have been found out when the coaxial probe was used in the frequency region from 0.5 GHz to 3 GHz. The composition of metakaolin specimens and the matrix low moisture make it difficult to measure the conductivity of relatively big size specimens. Capacitance measurements using short electrode distances appear to be more convenient.

## Acknowledgments

This paper has been worked out under the project GAČR No. 19-11516S supported by Czech Science Foundation.

**References**

[1] Han B, Ding S and Yu X 2015 *A review, Measurement* **59** pp 110–128
[2] Hou Z, Li Z and Wang J 2007 J *Wuhan Uni. Technol.-Mater. Sci. Ed.* **22** pp 346–349
[3] Guan H, Liu S, Duan Y and Cheng J 2006 *Cem. Concr. Compos.* **28** pp 468–474
[4] Fernandez-Jimenez A 1999 *Cement Concrete Res.* **29** p 1313
[5] Saafi M, Andrew K, Mcghon D, Taylor S, Rahman M, Yang S, Zhou X and Tang P L 2013 *Construction and Building Materials* **49** pp 46–55
[6] Ding Y, Chen Z, Han Z, Zhang Y and Pacheco-Torgal F 2013 *Constr. Build. Mater.* **43** p 233
[7] Rovnanik P, Kusak I, Bayer P, Schmid P and Fiala L 2019 *Cement and Concrete Research* **118 No. 1** pp 84–91
[8] Cabeza M, Merino P, Miranda A, Novoa X R and Sanchez I 2002 *Cement and Concrete Research* **32** pp 881–891
[9] Li GY, Wang P M and Zhao X H 2007 *Cem. Concr. Compos.* **29** pp 377–382
[10] Kusak I, Lunak M and Schauer P 2013 *Appl. Mech. Mater.* **248** p 370

IOP Conf. Series: Materials Science and Engineering 1039 (2021) 012021 doi:10.1088/1757-899X/1039/1/012021

# Testing methods of various materials for development of cement composites with low electrical resistance

**J Melichar[1], V Cerny[1], R Drochytka[1], P Steffan[2], P Dohnalek[3]**

[1] Institute of Technology of Building Materials and Components, Faculty of Civil Engineering, Brno University of Technology, Veveri 95, 602 00, Brno
[2] Department of Microelectronics, Faculty of Electrical Engineering and Communication, Brno University of Technology, Technická 3058/10, 616 00, Brno
[3] BETONCONSULT, s.r.o., V Rovinách 609/123, 140 00, Praha 4

Email: melichar.j@fce.vutbr.cz

**Abstract.** Our entire planet is filled by communication routes, and as a result of the evolving information technologies, their numbers will continue to grow in the future. Demand for their reliability and hence the importance of protecting related technologies from all kinds of interference is increasing as well. Therefore, it is necessary to develop new materials in order to protect these routes against negative influences of the environment such as the atmospheric discharge. The main aim of the research is to develop and verify the properties of the cement composite that can be applied in systems of protection of building structures against lightning strike and overvoltage. The first step in the development of this material was the selection of suitable feedstocks, which by their presence in the newly developed material will increase its electrical conductivity respectively reduce its electrical impedance. For this purpose, a spectrum of potential raw materials with a high content of metals and organic carbon was chosen. By means of a suitably selected set of laboratory methods which consisted of determination of specific surface, impedance and total organic carbon content (TOC), materials with the most suitable properties were selected and their parameters determined also in a cement matrix. In order to realize this goal specimens with each conducive material as filler were created with incorporated copper electrodes. Impedance has been significantly reduced compared to the reference samples. The lowest values in the tens of ohms were obtained from samples containing carbon grit 0.5–4.0. This fact proves that the tested fillers can be used in order to produce electrically conductive cement composites.

## 1. Introduction

In the vast majority of industries, environmental awareness has recently taken place as a result of changing waste management regulations. By 2020, the recycling rate should have increased to 50%. In 2024, an amendment to the Act, passed in 2014, will come into force, banning the storage of municipal and recyclable waste. Therefore, it is not only necessary to reduce the production of unnecessarily generated waste, but also to invent recovery for waste substances generated as by-products of some production and energy processes. Search for applications for waste products and the use of secondary raw materials is an even more important task now than ever. [1]

The current trend is the development of modern composite materials, the aim of which is to expand the number of functions that the materials can simultaneously perform. The appropriately chosen secondary raw materials could play an important role in the development of electrically conductive composites, which could not only significantly shift the area of pre-stress protection of buildings against lightning and overvoltage, but also its production should mean lower environmental burden and economic demands. [2]

Lightning and surge protection systems for buildings can be extended by use of the special materials mentioned above. Even a completely new functioning lightning and surge protection system based exclusively on the use of advanced composite materials can be developed. The system could not only be more functional in terms of lightning discharge, but also economically and environmentally acceptable. [3]

The aim of the research described in this paper is to develop new silicate composites capable of conducting electrical current to such extent that it will be possible to apply them in lightning protection or overvoltage protection systems due to the use of suitable electrically conductive fillers mainly from secondary sources. The subject of this article is the study of raw materials with the potential of the above mentioned parameters.

Resistivity of metals is in the values of $10^{-6}$ ohm·cm$^{-1}$ (silver, copper, aluminium, steel). The conductive composite material is prepared by admixing fine conductive particles into the cement matrix. It has been found that the conductivity of the composite increases sharply when a certain filler concentration value is reached, at which moment each particle in the matrix contacts at least two other adjacent particles, thereby forming a three-dimensional network in the matrix. This critical value is called the percolation threshold and is often expressed in the literature as the volume fraction of filler in the matrix. By increasing the filler concentration further above the critical limit, the resistivity does not decrease in proportion to the filler addition. [4]

As a possible filler for increasing the electrical conductivity of silicate building materials has been tested various materials, such as carbon fibres and steel shavings. [1] In other research graphite powder, steel powder and aluminium chip or steel fibres and graphites were tested in order to increase the electrical conductivity of concrete. [2, 3]

The percolation threshold also depends on the shape of the filler particles. An important aspect is the aspect ratio. Fibres or flakes with a large length to diameter (thickness) ratio have a much lower percolation threshold (less than 6 to 8 vol. %). The electrical conductivity can be reduced by an oxide layer covering some metals such as copper or aluminium. [5]

The addition of conductive or semiconductive particles to the non-conducting matrix affects the electrical properties in proportion to the amount, proximity and shape of the particles. The following three situations may occur with the additive particles:

- The particles do not touch – the particles are isolated from each other, the conductivity of the mixture changes only slightly or not at all. The composite material remains an insulator / non-conductor, but its dielectric properties vary significantly.
- Particles are very close to each other – electrons can jump the gaps between particles and allow the passage of electric current. The possibility of electron jump across the gap under a given (electrical) voltage increases exponentially with decreasing gap width between particles.
- Particles are touching – under these conditions the composite conducts electrical current through a network of conductive filler particles. The conductivity mechanism of the filler is applied here. Metal filler composites exhibit band conductivity. The band and flashover conductivity can be distinguished according to the AC or DC current behavior of the composite.

## 2. Materials and methods

The initial step in solving the issue was selection of a potentially appropriate raw material base. It contains both a binder component (Portland cement CEM I 42.5) and filler components. The filler components were expected to increase the electrical conductivity of the resulting mixtures. Therefore, it is important to conduct analysis and tests to determine the basic properties of the fillers under consideration.

Among the potentially suitable electrically conductive fillers were selected three types of carbon grit (UD N 0-2, UD N 0.5-4.0 and UD S 0.4-4.0) pictured in figure 1, solid salts and solutions containing heavy metals (PSK), Cooling Belt Dusts (OCP), Zirconium Waste Tablets ($ZrO_2$), Wood Pellet Pyrolysis

Waste (OPD) and as potentially suitable additives soot CHEZACARB and Carbon Black (CABOT) pictured in figure 2.

**Figure 1.** Three types of carbon grits (left to right: S 0.4–4.0; N 0–2; N 0.5–4,0).

**Figure 2.** Other types of potentially suitable fillers and additives (a – PSK, b – OCP, c – $ZrO_2$, d – OPD, e – CHEZACARB, f – CABOT).

In order to determine the required parameters and to choose the most suitable raw materials, a spectrum of laboratory methods was developed consisting of the determination of electrical conductivity (or rather impedance), specific surface area and total organic carbon content (TOC method).

In order to determine the electrical impedance, the sample was firstly loaded in the TRYSTOM press with a force of 10 kN (16 MPa). Impedance spectroscopy was measured for each pressure in the frequency range 1 MHz – 0.1 Hz and for potential values of 1, 10, 50 and 100 mV. The impedance value is read at 1 kHz. The thickness of the sample is calculated by first measuring the distance of the plates of the press with inserted measuring piston without the measured material, and then measuring the distance of the plates of the press with inserted measuring piston in which the measured material is already placed. From the difference of these distances, the thickness of the sample is determined, and the specific conductivity of the material is calculated from the measured impedance values. [6] The realization of testing method is pictured in figure 3.

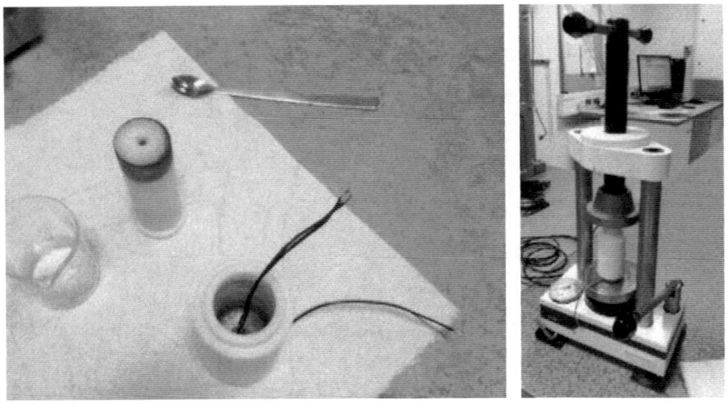

**Figure 3.** Procedure of impedance determination of individual raw materials.

The specific surface of the raw materials was determined using an AccuPyc II 1340 helium pycnometer. It is a very accurate method of measuring by varying the volume of helium (or other gas). [7]

The method for determining the total organic carbon content was based on the thermal and catalytic oxidation of organically bound carbon to $CO_2$, which was analytically determined, most often by analysis in the infrared region. [8]

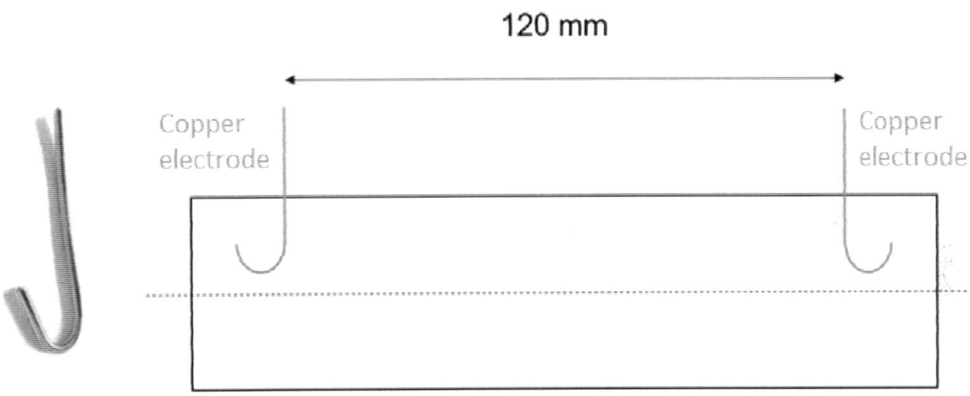


**Figure 4.** Diagram of placement of copper electrodes in the sample.

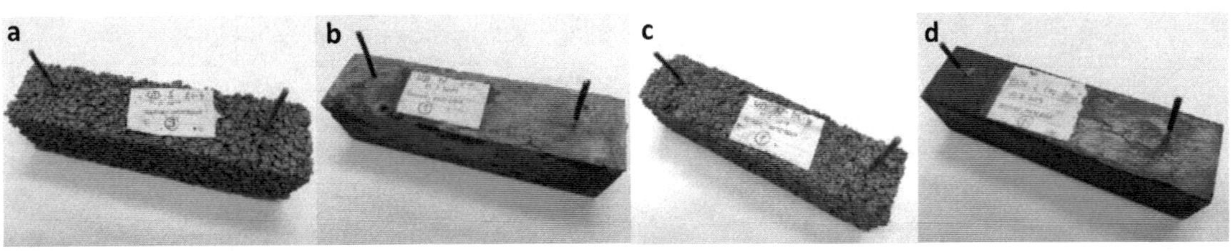


**Figure 5.** Test samples with four selected fillers (a – UD S 0.4–4.0; b – UD N 0-2; c – UD N 0.5–4,0; d – OCP).

### 3. Results and discussion

From the obtained results it is evident that the fillers PSK, ZrO2 and OPD exhibited insufficient impedance values and will not be considered in the further research. Filler OCP and all three types of carbon grit reached impedance values lower than 0.5 Ω and therefore their properties in cement matrix were further tested. For each type of filler, three beams of 40 × 40 × 160 mm were built with built-in electrodes as shown in figure 4. The composition of the mixture was 30% Portland CEM I 42.5 and 70% of the individual filler. Three test specimens (figure 5) were prepared from each material and their impedance determined. Both types of soot tested also appear to be perspective for further research. Table 1 shows the results of laboratory tests.

**Table 1.** Results of analysis of tested fillers.

| Filler | Specific surface [cm$^2$/g] | Impedance [Ω] (1 kH. 10 kN) | TOC [%] |
|---|---|---|---|
| UD S 0.4–4.0 | 230 | 0.002 | 97.12 |
| UD N 0–2 | 520 | 0.31 | 98.10 |
| UD N 0.5–4.0 | 220 | 0.009 | 98.21 |
| PSK | 770 | 12000000 | 0.81 |
| OCP | 2 560 | 0.38 | 1.41 |
| $ZrO_2$ | immeasurable | 1100000 | 0.0033 |
| OPD | 740 | 280.0 | 61.37 |
| CABOT | immeasurable | 0.003 | 99.20 |
| CHEZACARB | immeasurable | 0.022 | 98.76 |

unsatisfactory

During the measurements, the influence of the AC current frequency and the effect of the magnitude of the measuring voltage were monitored so that the magnitude of the measuring voltage and current is not on the limits of measurability. The impedance spectroscopy graph proved that at 1 kHz the impedance stabilized and could be considered steady. At this frequency, the real and imaginary components of the impedance have already stabilized. The measurement was carried out using an Agilent E4980 RLC meter (figure 6).

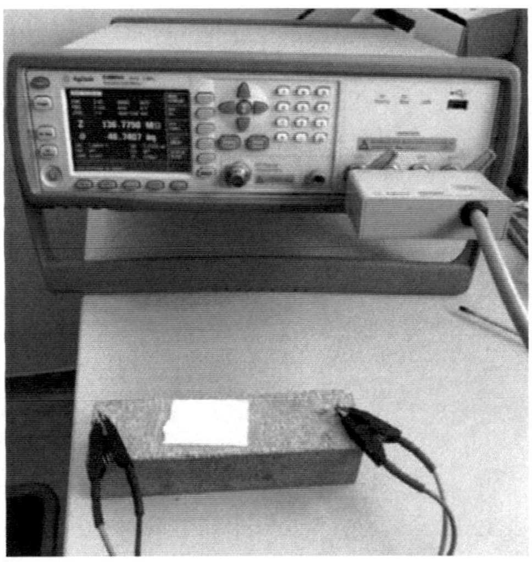

**Figure 6.** Impedance measurement by device Agilent LCR E4980A.

The resulting impedance values of the individual mixtures after drying the bodies to constant weight and stabilizing the obtained values are given in table 2 below.

**Table 2.** Results of impedance determination of test specimens.

| Mixture | Impedance [Ω] |
|---|---|
| UD S 0.4–4.0 | 140.8 |
| UD N 0–2 | $9 \cdot 10^6$ |
| UD N 0.5–4.0 | 31.9 |
| OCP | $272.3 \cdot 10^3$ |

## 4. Conclusion

In the development of new silicate composites capable of conducting electric current, a spectrum of materials has been assembled, mainly from secondary sources, which should help this goal. The materials were chosen on the basis of the high proportions of metals or organic carbon in their structure. Their parameters were determined by laboratory methods and four fillers with the best parameters were chosen for further testing. Suitable fillers were selected mainly on the basis of their electrical impedance values. The impedance values of the selected materials were then verified also in the cement matrix. The final results show that the resistivity, respectively electrical impedance, has been reduced to very low values, thus proving that by usage of the tested fillers (especially UD S 0.4–4.0 and UD N 0.5–4.0) it is possible to produce electrically conductive cement composites.

## Acknowledgments

This paper was elaborated with the financial support from funds of the state budget through the Ministry of Industry and Trade within the project FV30239 - "Advanced Earthing Enhancing Materials in the Lightning and Overvoltage Protection System".

## References

[1] Malakooti A, Theh W S, Sadati S, Ceylan H, Kim S, Mina M, Cetin K and Taylor P C 2020 Design and Full-scale Implementation of the Largest Operational Electrically Conductive Concrete Heated Pavement System *Construction and Building Materials* Volume **255** ISSN 0950-0618

[2] Shishegaran A, Daneshpajoh F, Taghavizade H and Mirvalad S 2020 Developing conductive concrete containing wire rope and steel powder wastes for route deicing *Construction and Building Materials* Volume **232** ISSN 0950-0618

[3] Rao R, Fu J, Chan Y, Tuan C Y and Liu C 2018 Steel fiber confined graphite concrete for pavement deicing *Composites Part B: Engineering* Volume **155** pp 187–196 ISSN 1359-8368

[4] Berrocal C G, Hornbostel K, Geiker M R, Löfgren I, Lundgren K and Bekas D G 2018 Electrical resistivity measurements in steel fibre reinforced cementitious materials *Cement and Concrete Composites* Volume **89** pp 216–229 ISSN 0958-9465

[5] Banthia N, Djeridane S and Pigeon M 1992 Electrical resistivity of carbon and steel micro-fiber reinforced cements *Cement and Concrete Research* Volume **22** Issue 5 pp 804–814 ISSN 0008-8846

[6] Han B, Zhang K, Yu X, Kwon E and Ou J 2012 Electrical characteristics and pressure-sensitive response measurements of carboxyl MWNT/cement composites *Cement and Concrete Composites* Volume **34** Issue 6 pp 794–800 ISSN 0958-9465

[7] ČSN EN 196-6:2010 Cement testing methods: Part 6: Determination of fineness of grinding (Prague: CNI)

[8] ČSN EN 15936 Sludges, treated organic waste, soils and wastes - Determination of total organic carbon (TOC) by dry incineration (Prague: CNI)

IOP Conf. Series: Materials Science and Engineering 1039 (2021) 012022 doi:10.1088/1757-899X/1039/1/012022

# Correction in determination of specific activity of radionuclides by gamma spectrometry in building materials

**V Polouckova**[1]

[1] Research Institute for Building Materials, Hnevkovskeho 30/65, Brno 617 00, Czech Republic

Email: radionuklidy@vustah.cz

**Abstract**. For more accurate determination of specific activities of radionuclides by gamma spectrometry in materials, it is necessary to take into account two important factors. These are the effect of true coincidences and self-absorption of gamma radiation directly in the measured sample. The corrections for these factors are determined for the specific measurement geometry defined by the detector, the measuring vessel and their relative positions. Corrections are calculated by Monte Carlo simulation. The results show a significant effect of material bulk density on the correction values.

## 1. Introduction

The content of natural radionuclides in building and solid materials is determined by gamma spectrometry [1, 2]. The detection device of the spectrometric assembly registers gamma radiation, which is emitted from the material during the radioactive decay of the contained radionuclides. The energy of the detected gamma radiation is characteristic for individual radionuclides. The output is an energy spectrum, from which it is possible to determine the specific activities of the radionuclides contained in the material after calibration of the measuring device.

Some radionuclides have more complicated decay pattern and they can emit gamma radiation of more energies over a very short period of time, which is detected by the device as a single event with the value of the sum of their energies. Therefore, it is appropriate to make corrections for true coincidences in the activities of these radionuclides [3, 4].

To quantify the specific activity of individual radionuclides, it is necessary to perform an efficient calibration of the spectrometer with a certified standard [5]. The different composition of the standard with which the spectrometer is calibrated and of the measured samples affects the values of the measured specific activity of radionuclides. A correction is necessary to obtain the correct value. The specific activities determined for the individual gamma energies of the given radionuclides are corrected. At lower energies, the correction can reach tens to hundreds of percent [2].

The corrected specific activity $A_{cor}$ is obtained by multiplying the measured specific activity $A$ by corrections for self-absorption $K_s$ and by correction for true coincidences $K_c$.

$$A_{cor} = A \cdot K_s \cdot K_c \tag{1}$$

The size of the corrections is influenced mainly by material composition, material bulk density, the energy of the emitted gamma radiation and the geometry of the measurement.

For the normal operation of a laboratory that determines the specific activities of several radionuclides in many different materials, it is necessary to find a fast, inexpensive and sufficiently accurate method for determining corrections. According to [6] and [7], where the authors compared several methods, the Monte Carlo simulation computational method seems to be a suitable approach.

## 2. Methods and materials

### 2.1. EFFTRAN

The program EFFTRAN [8] was chosen to calculate true coincidence summing correction and self-absorption correction. The software calculates self-absorption correction and true coincidence summing correction using Monte Carlo simulation.

The input data is the material composition of the detector, measuring vessel and measured sample. It is necessary to load the material bulk density into the program, the contained elements or compounds in the form of chemical formulas (e.g. $SiO_2$, Ge, $H_2O$) and their representation in the material in wt. %. The program creates a file with specified parameters and a file with calculated material attenuation coefficient for different energies of gamma radiation [3, 9, 10].

Furthermore, the detector parameters and measuring geometry are defined. The detector is characterized by dimensions and material composition, which is stated in the documents of the detector producer. The geometry of the measurement is determined by material and dimensions of the measuring vessel, the mutual position of the detector and the sample. Polypropylene cylindrical measuring vessel with a volume of 0.5 l was placed directly on the detector head.

Other parameters for calculating the attenuation of gamma radiation in the sample and self-absorption correction for selected radiation energies are the material of the calibration standard, the material of the measured sample and the efficiency of the detector [3].

While maintaining the same measurement geometry, the relation applies to the correction for self-absorption $K_s$ as follows:

$$K_s = \frac{\eta_{et}}{\eta_{sa}}, \tag{2}$$

where is:
$\eta_{et}$ ... detector efficiency for standard,
$\eta_{sa}$ ... efficiency for the measured sample.

### 2.2. Standard

When verifying the spectrometer, an efficiency calibration is performed with an activity standard of CBSS 2 type. The radioactive material is homogeneously dispersed in silicone rubber with a bulk density of 0.98 g/cm$^3$. The material composition is given in table 1.

**Table 1.** Composition of standard.

| composition | Si | C | O | H |
|---|---|---|---|---|
| wt. % | 37.9 | 32.4 | 21.6 | 8.2 |

### 2.3. Definition of materials

The most common building materials used in laboratory practice were selected. The material composition was found in the literature or chemistry analysis was performed. The chemical composition and bulk density of selected materials are given in table 2. The meaning of bulk density in this article is the ratio of the sample weight and the volume of measuring vessel. Samples of building materials are dried and crushed under 8 mm.

**Table 2.** Composition of building materials.

| | plaster | cement | concrete | slag | fly-ash | brick |
|---|---|---|---|---|---|---|
| | | | [wt. %] | | | |
| **CaO** | 62 | 64 | 11 | 5 | 2 | 10 |
| **$SiO_2$** | 16 | 20 | 63 | 52 | 53 | 58 |
| **$Al_2O_3$** | 4 | 5 | 13 | 23 | 30 | 15 |
| **$SO_3$** | 3 | 3 | 1 | 2 | 1 | 2 |
| **$Fe_2O_3$** | 3 | 3 | 3 | 11 | 9 | 5 |
| **$H_2O$** | 12 | 4 | 2 | - | 1 | 1 |
| **MgO** | - | 1 | 2 | 2 | 2 | 4 |
| **$K_2O$** | - | - | 3 | 4 | 2 | 3 |
| **$Na_2O$** | - | - | 2 | 2 | - | 2 |
| **bulk density [g/cm$^3$]** | 0.8–1.8 | 1.5–1.8 | 1.1–2.4 | 0.7–2.2 | 0.7–1.6 | 1.0–1.6 |

*2.4. Parameters*

Natural radionuclides emit gamma radiation in the range of energy values 46–1764 keV during their decay. The standard analysis of building materials determines the specific activity of the three radionuclides $^{40}$K, $^{226}$Ra and $^{228}$Th and the activity concentration index I calculated from them [1, 11]. The specific activity of $^{40}$K is determined directly from one energy line. The specific activity of radionuclides $^{226}$Ra and $^{228}$Th are calculated as weighted activities from the three energy lines of their decay daughter products, with which they are in radionuclide equilibrium. Table 3 shows the line energies for determining the specific activity of radionuclides [2, 4].

**Table 3.** Gamma-ray energies in building materials.

| Energy [keV] | Radionuclide | Source of radiation |
|---|---|---|
| 238.63 | $^{228}$Th | $^{212}$Pb |
| 295.21 | $^{226}$Ra | $^{214}$Pb |
| 351.93 | $^{226}$Ra | $^{214}$Pb |
| 583.19 | $^{228}$Th | $^{208}$Tl |
| 609.31 | $^{226}$Ra | $^{214}$Bi |
| 727.33 | $^{228}$Th | $^{212}$Bi |
| 1460.80 | $^{40}$K | $^{40}$K |

Activity concentration index I is calculated according to the formula [1]:

$$I = \frac{A_K}{3000} + \frac{A_{Ra}}{300} + \frac{A_{Th}}{200}, \quad (3)$$

where is:

$A_K$ … specific activity of $^{40}$K,

$A_{Ra}$ … specific activity of $^{226}$Ra,

$A_{Th}$ … specific activity of $^{228}$Th.

## 3. Results and Discussion

### *3.1. Self-absorption correction*

The size of self-absorption correction $K_s$ is influenced mainly by material composition, material bulk density and the energy of attenuated gamma radiation [3, 10].

A concrete sample was selected to monitor the effect of bulk density on the correction size. The concrete composition in the calculations was the same but the value of the material bulk density changed. The shape of the curves of the dependence of $K_s$ on the radiation energy, see figure 1, is identical with the shape obtained by experimental methods [6] and simulation [10]. The value of the self-absorption correction increases steeply for low energies below 100 keV [9].

The linear dependence of $K_s$ on the bulk density is illustrated in figure 2. The slope of the line increases with decreasing of the radiation energy. There is a steep increase in the values of the slopes for energies below 100 keV. The value of $K_s$ increases with increasing bulk density [12–15]. This can be explained by the fact that lower energy gamma radiation is significantly absorbed in the sample and therefore is not registered by the detector. Likewise, the materials with a higher bulk density significantly attenuate gamma radiation.

The bulk density of different kinds of building material varies a lot. However, analysing a large number of samples, it was possible to choose different types of building materials with the same bulk density (as it is mention above, bulk density in this article means the ratio of the sample weight and the volume of measuring vessel). The bulk density was chosen 1.6 g/cm$^3$ as it fits to most of measured types of building materials. For some materials, e.g. fly ash, this is a limit value.

For building materials with the same bulk density and different compositions, the $K_s$ values differ within a one percent difference to energy values around 150 keV. Below an energy of 100 keV, a significant influence of different composition of materials starts to be seen. The self-absorption correction increases significantly with decreasing value of gamma radiation energy, see figure 3. Bulk density of 1.6 g/cm$^3$ was chosen to illustrate the effect of the composition and energy of the emitted radiation on the correction coefficients.

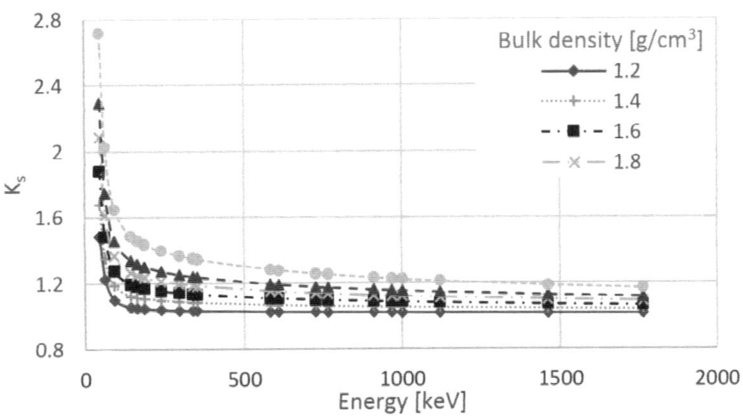


**Figure 1.** Correction $K_s$ of concrete with different bulk density.

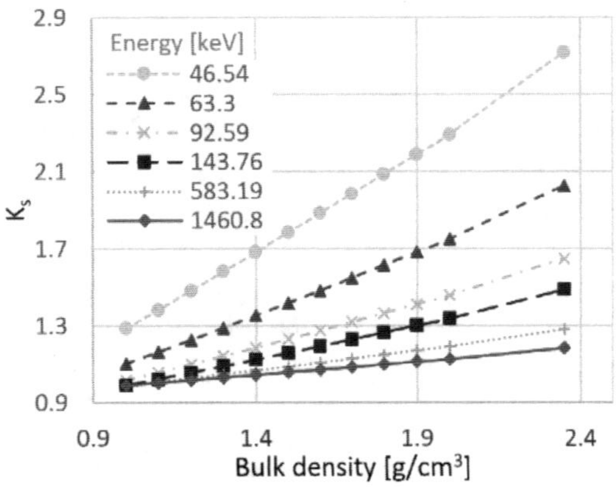


**Figure 2.** Correction $K_s$ of concrete.

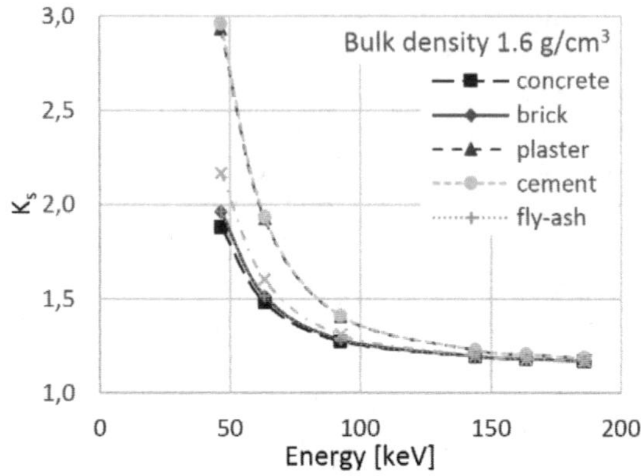


**Figure 3.** Correction $K_s$ of materials with different composition.

### 3.2. True coincidence summing correction

The values of true coincidence summing correction $K_c$ for different building materials with the same bulk density of 1.6 g/cm$^3$ are practically unaffected by the material composition for the mentioned representatives of building materials or only very weakly, see table 4.

**Table 4.** Correction $K_c$ of building materials with different composition and same bulk density 1.6 g/cm$^3$.

| Energy [keV] | concrete | cement | brick | plaster | fly-ash | slag |
|---|---|---|---|---|---|---|
| **238.63** | 1 | 1 | 1 | 1 | 1 | 1 |
| **295.21** | 1 | 1 | 1 | 1 | 1 | 1 |
| **351.93** | 1.001 | 1.001 | 1.001 | 1.001 | 1.001 | 1.001 |
| **583.19** | 1.093 | 1.093 | 1.093 | 1.093 | 1.093 | 1.093 |
| **609.31** | 1.077 | 1.077 | 1.077 | 1.077 | 1.077 | 1.076 |
| **727.33** | 1.024 | 1.024 | 1.024 | 1.024 | 1.024 | 1.024 |
| **1460.80** | 1 | 1 | 1 | 1 | 1 | 1 |

**Table 5.** Correction $K_c$ of concrete with different bulk density.

| Energy [keV] | 1.0 | 1.2 | 1.3 | 1.4 | 1.5 | 1.6 [g/cm$^3$] | 1.7 | 1.8 | 1.9 | 2.0 | 2.4 |
|---|---|---|---|---|---|---|---|---|---|---|---|
| **351.93** | 0.999 | 1.001 | 1.001 | 1.001 | 1.001 | 1.001 | 1.001 | 1.001 | 1.001 | 1.001 | 1.000 |
| **583.19** | 1.082 | 1.086 | 1.088 | 1.09 | 1.091 | 1.093 | 1.095 | 1.097 | 1.099 | 1.101 | 1.110 |
| **609.31** | 1.067 | 1.07 | 1.072 | 1.074 | 1.075 | 1.077 | 1.079 | 1.08 | 1.082 | 1.084 | 1.091 |
| **727.33** | 1.021 | 1.022 | 1.022 | 1.023 | 1.023 | 1.024 | 1.024 | 1.025 | 1.025 | 1.026 | 1.028 |

True coincidence summing correction $K_c$ for selected most commonly measured building material, i.e. concrete, with different bulk densities is given in table 5. It is not necessary to perform true coincidence summing correction for gamma energies 238.63 keV, 295.21 keV and 1460.80 keV. A more significant effect of material bulk density on $Kc$ value is manifested especially at an energy of 583.19 keV and 609.31 keV. Within the stated interval of real values of concrete bulk density, the value of $K_c$ increased by 0.015 for the line 583.19 keV and by 0.014 for the line 609.31 keV as the bulk density increased from 1.2 to 2.0 g/m$^3$. For the 727.33 keV line, the increase was lower, only by 0.004. As the bulk density of the material increases, the value of $K_c$ also increases, see figure 4. This is then reflected in the values of the corrected specific activities calculated as a weighted average of the specific activities determined from different energy lines.

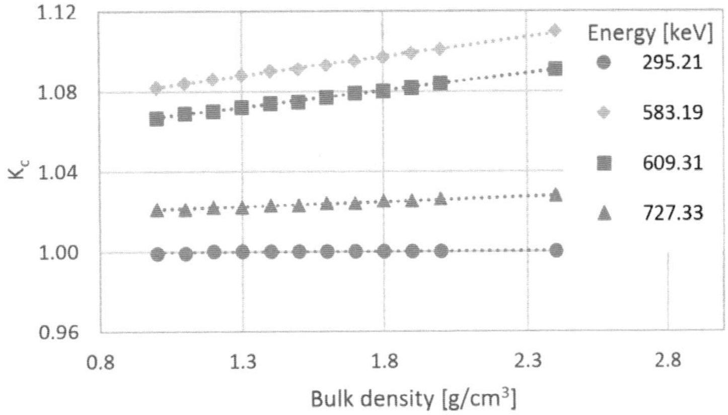


**Figure 4.** Correction $K_c$ of concrete.

### 3.3. *Influence of correction on the correctness of determined parameters*

The relative deviation of the corrected values from the measured value was chosen as a quantity illustrating the effect of corrections on the determined parameters. The relative deviations δ of specific activities of radionuclides and activity concentration index are calculated according to relation (4):

$$\delta = \left(\frac{A_{cor,i}}{A_{mea,i}} - 1\right) \cdot 100\ \%, \tag{4}$$

where is:
$\delta$ ......... deviation,
$A_{cor,i}$ ... corrected parameter,
$A_{mea,i}$ ... measured parameter,
$i$ .........specific activity of $^{40}K$, $^{226}Ra$, $^{228}Th$ and I.

The effect of the corrections of specific activities of radionuclides and activity concentration index for different real materials of different composition and bulk density is shown in figure 5. For materials with higher bulk density, a higher deviation of the corrected values from the measured ones is evident.

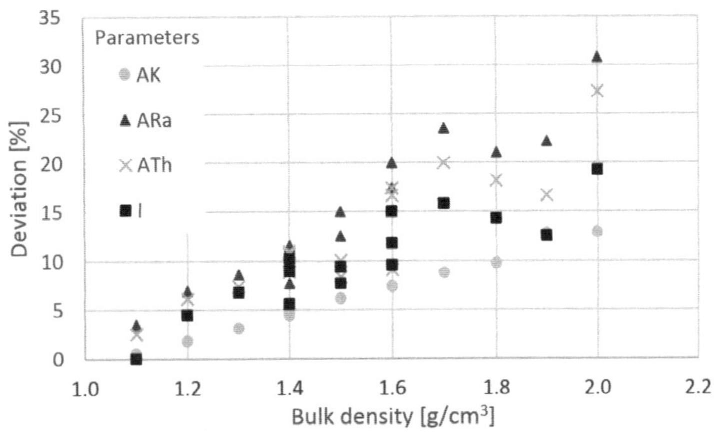


**Figure 5.** Deviation of parameters of real building materials.

## 4. Conclusion

The results show that the sample composition is not a determining factor for true coincidence summing correction and for self-absorption correction for radiation energies above 100 keV. The advantage of this method is that for the energies of radionuclides used in gamma spectrometry of building materials it is not necessary to analyse the material composition, it is enough to know the bulk density of the material. The EFFTRAN software is a fast and efficient way to obtain the self-absorption coefficient for an energy of 100–1700 keV.

For photon energies lower than 100 keV, which are used to measure the activities of radionuclides in radioactive substances like waste from NORM workplaces, e.g. $^{210}$Pb with energy 46.54 keV, the influence of the material composition of the sample on the size of the self-absorption coefficient is significantly applied. At such low photon energies, EFFTRAN can only be used with knowledge of the chemical composition of the sample. If the exact composition is not known, it is necessary to use another method.

Making true coincidence summing corrections and self-absorption corrections significantly contributes to determining the correct values of specific activities and activity concentration index. For the measurement geometry used, the results show that when no corrections are performed, especially at higher material densities, the specific activities are underestimated by up to 30%.

**Acknowledgements**

This paper was supported by the Ministry of Industry and Trade of the Czech Republic within the Institutional Support for Long-term Development of a Research Organization.

**References**

[1] *Doporučení, měření a hodnocení obsahu přírodních radionuklidů ve stavebním materiálu* 2017 (Praha: Státní úřad pro jadernou bezpečnost)

[2] *Doporučení, měření a hodnocení obsahu přírodních radionuklidů v radioaktivní látce uvolňované z pracoviště s možností zvýšeného ozáření z přírodního zdroje záření podle § 95 odst. 1 písm. b) atomového zákona* 2017 (Praha: Státní úřad pro jadernou bezpečnost)

[3] Gilmore G 2008 *Practical Gamma-ray Spectrometry* (Chichester: John Wiley & sons.) ISBN 978-0-470-86196-7

[4] Ordóñeza J, Gallardob S, Ortizc J and Martorellac S 2019 Coincidence summing correction factors for 238U and 232Th decay series using the Monte Carlo method *Radiation Physics and Chemistry*Volume **155** pp 244–247

[5] Poloučková V, Jahelková D, Persaň V and Sedlák J 2019 *Vliv změny složení betonové směsi na obsah přírodních radionuklidů* (Online: https://www.tzb-info.cz/14537-vliv-zmeny-slozeni-betonove-smesi-na-obsah-prirodnich-radionuklidu)

[6] Svobodová T, Hýža M, Rulík P and Vlček J 2015 Porovnání postupů pro výpočet korekce na samoabsorpci fotonů *Sborník XXXVII. Dny radiační ochrany* (Praha) p 102

[7] Možnar R and Johnová K 2018 Současné metody korekcí na samoabsorpci a pravé koincidence v laboratorní gama spektrometrii *Sborník XL. Dny radiační ochrany* (Praha) p 138

[8] Vidmar T 2005 EFFTRAN - a Monte Carlo efficiency transfer cod efor gamma-ray spektrometry *Nuclear Instruments and Methods* A **550** pp 603–608

[9] Carrazana González J, Cornejo Díaz N, Jurado Vargas M and Capote Ferrera E 2010 The effect of source chemical composition on the self-attenuation corrections for low-energy gamma-rays in soil samples *Applied Radiation and Isotopes* **68**(2) pp 360–363

[10] Johnová K 2016 Self-absorption corrections based on Monte Carlo simulations *Acta Polytechnica CTU Proceedings* A**4** pp 27–32

[11] Directive 2013/59/EUROATOM *Official Journal of the Eurepean Uniun* 2014

[12] Mostajaboddavati M, Hassanzadeh S, Faghihian H, Abdi M R and Kamali M 2006 Efficiency calibration and measurement of self-absorption correction for environmental gamma-spectroscopy of soil samples using Marinelli beaker *Journal of Radioanalytical and Nuclear*

*Chemistry* **268**(3) pp 539–544

[13] Vargas M Jurado, Timón A Fernández, Díaz N Cornejo and Sánchez D Pérez 2002 Monte Carlo simulation of the self-absorption corrections for natural samples in gamma-ray spectrometry *Applied Radiation and Isotopes* **57**(6) pp 893–898

[14] Jodłowski P 2006 Self-absorption correction in gamma-ray spectrometry of environmental samples – an overview of methods and correction values obtained for the selected geometries *NUKLEONIKA* **51**(Supplement 2) pp S21–S25

[15] Lotfi M, Mostajabodaavati S M and Sichani B T 2012 Study of self-absorption correction for environmental gamma-spectroscopy of soil samples using marinelli beaker *1st International Conference on Radiation and Dosimetry in Various Fields of Research, RAD Conference Proceedings* Volume 2012-April pp 343–346

# Instrumented dynamic tests of building materials

**M Sperl[1], Ľ Gajdos[1], Z Slizkova[1] and M Drdlova[2]**

[1] Institute of Theoretical and Applied Mechanics of the Czech Academy of Sciences, Prosecká 809/76, 190 00 Prague 9, Czech Republic
[2] Research Institute for Building Materials, Hnevkovskeho 65, 617 00 Brno, Czech Republic

Email: sperl@itam.cas.cz

**Abstract.** In some specific cases, the building structures can be exposed to extreme dynamic loads. Namely, it can be car crash, earthquake or the effect of explosion. For these cases, it is useful to know the dynamic mechanical properties of building materials. The paper presents two unique procedures for instrumented impact testing of building materials (concrete and sandstone). These tests may help in practice with selection of suitable materials for buildings and structures which may be threatened by dynamic effects of mechanical stresses. The method how to achieve a higher dynamic material stability is also presented.

## 1. Introduction

Besides the development of new progressive building materials and technologies aiming at improving physical properties of currently used materials it is also necessary to develop some experimental testing methods for characterization of further partial mechanical properties. Proper mechanical characterization leads to better understanding of material potentialities and enables application of building materials appropriately to particular structures. Regarding the mechanical properties characterizing the behaviour of materials under mechanical loading there are mainly considered static properties, fracture mechanical properties, fatigue characteristics and properties relating to dynamic loads [1, 2]. Currently, the dynamic properties of building materials are tested via various drop tests, split Hopkinson tests, Charpy tests, etc. The problem is that standardization of these tests for building materials is rather low and thus is lower also the global comparability of results [3] and [1].

Dynamic loading of building structures can happen in many cases. Mostly, they are extreme states which should be considered in the stage of design of a building structure according to its utilization and location. This involves buildings situated at seismically active regions where, upon strong earthquakes, vibrations of acceleration of up to 1100 $cm/s^2$ (it means around 1 g) can be experienced [4]. Another dynamic loading of a structure can be caused by an explosion (terrorism, industrial premises), e.g. the shock wave created by detonation of TNT has the velocity of 6,900 m/s [5]. Another frequent case of dynamic impact on structures is car crash. Better understanding of the behaviour of materials under dynamic load can thus help in design of structures, in particular the structures with enhanced dynamic impact resistance.

The authors of this paper have aimed at the verification of the functionality of two procedures of dynamic instrumented tests. The former dynamic test comes from the modification of the ISO [6] test which is used in particular for impact tensile tests of polymers. For this test, Charpy impact tester WPM with the capacity 300 J was modernized and modified including its instrumentation to enable impact tensile testing of reinforced concretes. The latter dynamic test was accomplished by means of a modern instrumented Charpy impact tester Instron Ceast 9050, namely in the mode of three-point bending impact using specimens from sandstone. Consolidated specimens as well as non-consolidated notched specimens were used. The test was motivated by the standards [7–10] which are used for plastics and metals. There is no relevant standard yet in the field of testing building materials which would specify the procedure of this test and define the size of specimens, for example for concrete, natural stone, etc. The lack of a standardized method for Charpy testing of cementitious composites has led to a great deal of variation in execution between authors.

## 2. Dynamic tensile test

### *2.1. Materials and preparation of specimens*

Fine-aggregate high-performance concrete was selected for pilot dynamic tensile tests as a standard. This type of concrete is compatible with both fibres selected; owing to the limited grain size and excellent workability of the material due to additives, good dispersion of the fibres can be achieved. Fine $SiO_2$ sand with 0–1 mm grain, CEM 52.5R cement and water were the main components of the concrete mix. Glenium422 super-plasticizer (produced by BASF) was added to achieve good workability with a low water/binder ratio. Elkem 940U silica fume (also produced by BASF), with a typical particle size of 100–500 nm, was used to create an optimized particle packing density, and also for its pozzolanic properties. The mix proportions of the concrete are described in table 1. Two different types of fibre reinforcement were selected to create specimens with different tensile strength to verify easily the validity of results of the test method. During the pilot testing, only two samples were used per one fiber type. A statistically relevant sets of samples are now being prepared in modified moulds.

**Table 1.** Description of the mix.

| Designation | CEM52.5 R [kg·m$^{-3}$] | Fine aggregate [kg·m$^{-3}$] | Silica fume [kg·m$^{-3}$] | plasticizer [kg·m$^{-3}$] | Fibre content [Vol. %] | Fibre material | Length/diameter of the fibre (mm/μm) |
|---|---|---|---|---|---|---|---|
| A | 955 | 1050 | 115 | 15 | 0.5 | Aramid | 6.0/12 |
| S | 955 | 1050 | 115 | 15 | 0.7 | Steel | 6.0/150 |

Steel moulds with polymeric inserts were used to cast the specimens. The mixing procedure was as follows: first, the dry components (cement, sand and microsilica) were mixed together using a concrete mixer, and then the required quantity of water with plasticizer was added and mixed for 3 min. The fibres were incorporated at the end of the mixing process and were mixed for another 60 seconds to achieve good dispersion without any visible agglomerations. The mixes were placed in moulds treated with releasing agent and a vibration table was employed for better compaction of the mixture. The test specimens were demoulded after 24 h and cured for 28 days in water-filled curing ponds.

### *2.2. Testing procedure and results*

Modernization of the Charpy impact tester WPM with the capacity 300 J, including its instrumentation, consisted of the design of appropriate fixtures for clamping a specimen for tensile tests. For instrumentation the fixed chucked head was complemented by a suitable strain gauge transducer. The transducer is based on the shear principle and is used for force recording in the impact process. It consists of the full bridge $4 \times 350\ \Omega$ and its force range is limited by 80 kN. The sensitivity of the signal 1 mV/V corresponds to loading by horizontal force of 34.8 kN. This specific strain gauge bridge was further connected to the direct current strain gauge amplifier and the whole process of the impact test was monitored in – time by means of a digital oscilloscope connected to PC with a measuring software. The principle of the impact tensile test consists in translating kinetic energy of the impact tester hammer to the tensile specimen through the yoke. The yoke is freely set on the specimen and after the impact of the hammer it acts by tension on the specimen by means of its geometric configuration. The yoke is in a direct contact with the inside surface of one of the two heads of the specimen. The energy consumed for pushing-off the mass of the very yoke is subtracted from the result. After the specimen is broken

IOP Conf. Series: Materials Science and Engineering 1039 (2021) 012023 doi:10.1088/1757-899X/1039/1/012023

the yoke and a part of the specimen are repulsed to the direction of the movement of the hammer. For a better understanding, a scheme of the test and a set of photos from experiments are presented in figure 1. The principle is also described in the standard [6]. The impact speed of the Charpy hammer was 5.6 m/s with total impact energy capacity 7.5 J.

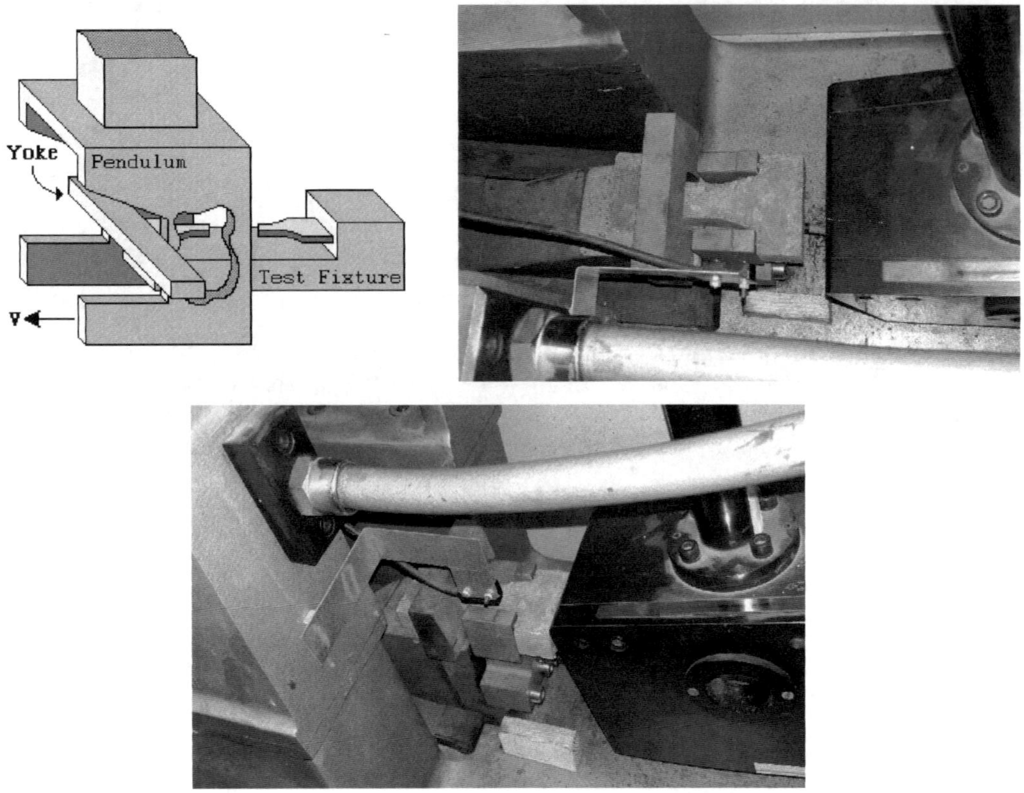


**Figure 1.** Test in tensile impact-strength mode.

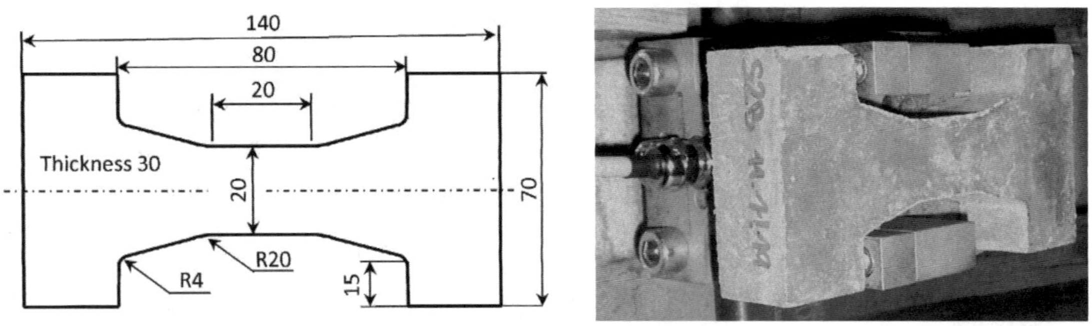


**Figure 2.** Tensile impact specimen.

The real shape of the tensile specimen is shown in figure 2. The results of the tests are (i) magnitudes of the energy needed for breaking specimens and (ii) records of the force – time dependence upon impact tension. All results are presented in table 2 and figure 3. Table 2 also summarizes the tensile peak forces recorded within the impact process.

**Table 2.** Results – Impact-strength test.

| Designation | Fibre material | Total fracture energy [J] | Peak force Fmax [kN] |
|---|---|---|---|
| S2B_1 | Steel | 120 | 76.1 |
| S2B_2 | Steel | 90 | 75.2 |
| A_1 | Aramid | 51 | 73.4 |
| A_2 | Aramid | 57 | 73.4 |

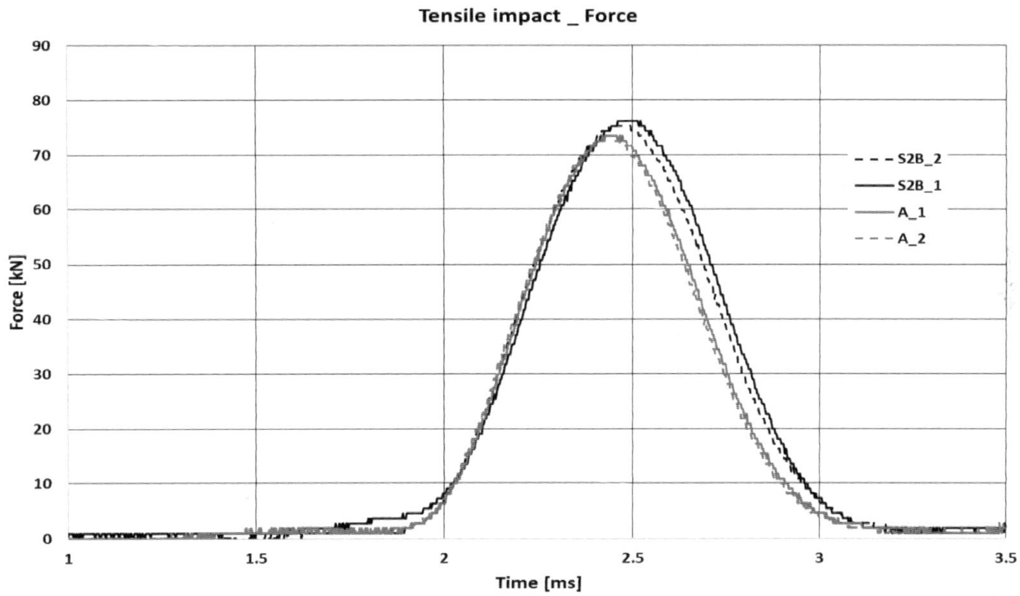


**Figure 3.** Comparison of dependences „Force – Time“ during dynamic tensile impact-strength tests.

It can be seen from the results of pilot tests that the test procedure is functional, however the shape of test specimens is not ideal. Due to a high sensitivity of concrete to stress concentration at notches the failure of specimens with aramid fibres took place at the transition from the grip holder to the measuring length. This problem did not occur when testing specimens with steel fibres. These specimens always failed at the beginning of the measuring length. Differences in fractures as well as the fracture surfaces are seen in figure 4.

Based on the experience obtained, the modification of the shape parameters of the test specimens will be performed. According to the nature of material and used test method, as an ideal shape a tensile type specimen with a zero-gauge length was selected. Further systematic research will continue with such prepared specimens.

Although the specimens with aramid fibres fractured in the locations with a significantly larger load - bearing cross section than that of specimens with steel fibres, it was found that their tensile resistance was lower than that for specimens with steel fibres. Both maximum tensile forces attained and energy consumed for dynamic tensile fracture of a specimen were lower in case of specimens reinforced with aramid fibres, thus the results of both obtained records are consistent. To be specific, the maximum average force for specimens with steel fibres was found to be higher by 2.3 kN and the consumed energy to be higher by 51 J on average. A zero equidistance of curves representing the „force – time“ dependence at their ascending part is worthy of remark. This attests to equal dynamic E modulus for both compositions of materials. This is caused more likely by the same composition of the concrete matrix and its volume as compared to the volume of fibres, especially in the case of aramid fibres.

The problem of the effect of volumetric part of fibres on the development of E modulus was dealt with by e.g. [13].

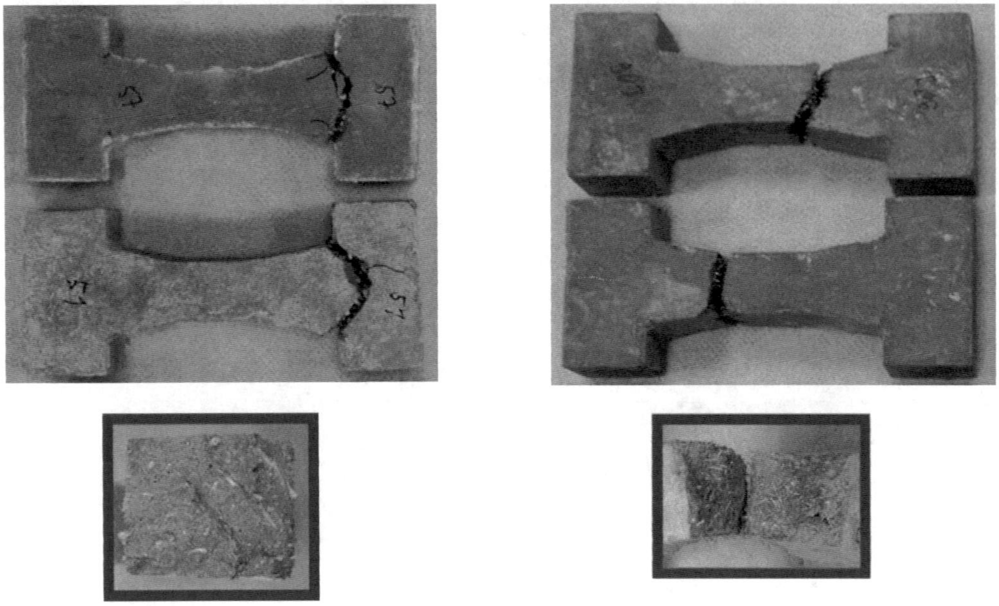

**Figure 4.** Left side: Specimens with aramid fibres and fracture surface,
Right side: Specimens with steel fibres and fracture surface.

## 3. Instrumented Charpy test for sandstone

### *3.1. Materials and preparation of specimens*

The stone used for testing was the arkose sandstone, quarried in Božanov (CR). The material can be characterized as a quasi-brittle inelastic silicate composite [11]. Test specimens were cut from the sandstone semiproduct with dimensions of 10 mm × 10 mm × 88 mm. An initiation notch 1 mm thick and 1.5 mm deep was made on one side of each specimen in the middle of the specimen length. The initiation notches were situated identically in the direction of sedimentation. The real shape of the Charpy specimen is shown in figure 5.

**Figure 5.** Charpy specimen.

The dynamic tests were accomplished by means of a modern instrumented Charpy impact tester Instron Ceast 9050 in the mode of three-point bending impact. Consolidated specimens as well as non-consolidated notched specimens were used. Stone specimens were treated by the stone strengthener, REMMERS KSE 300 E, with a silicic acid ester base. This product contained a special elasticised component and the possible effect of the component on strengthening sandstone was evaluated. The expected $SiO_2$ gel deposition rate for REMMERS KSE 300 E was around 30 w%. According to the producer information [12], the 'elasticised' stone strengthener could perform better in the case of stones with large pores or stones containing micro-cracks. The treatment of stone specimens with the liquid strengthening products was performed by placing the specimens in a dish, followed by gradual pouring the liquid up to 5 mm above the height of the liquid level (which represented the half of the specimen height). After the time period of two hours, the impregnated stone specimens were removed from the liquid product, the wet surface was shortly whipped using a paper and then the specimens were weighted to monitor the amount of the liquid soaked by stone. Before mechanical testing, all specimens were stored for 4 weeks at 20°C and 60% RH and during this time the chemical reaction responsible for the $SiO_2$ gel formation in stone pores occurred and the organic part of the product released. The weight gains after the four weeks of the storage were recorded for all specimens and the total weight increase in relation to the treatment was calculated. The obtained data are presented in Table 3. The measured product gel amount in the treated stone specimens (four weeks after the treatment) was 31.5 w% for the REMMERS KSE 300 E. This finding was in a good conformity with the technical data provided by manufacturer.

### *3.2. Testing procedure and results*

The test was motivated by the standards [7–10] which are used for plastics and metals. The test procedure is very similar to tensile impact test, with following differences: a specimen is loaded in the mode of three point bending impact, loading sensor is placed into the active part of the hammer and the shape of the hammer is modified. The impact speed of the Charpy hammer was 3.8 m/s with total impact energy capacity 7.5 J. The set of photos from experiments can be seen in figure 6. The span of supports was 60 mm [10].

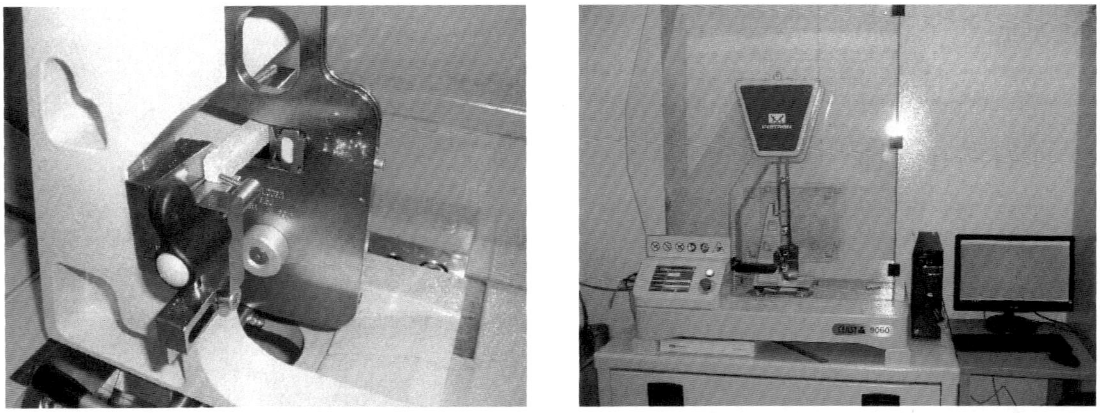

**Figure 6.** Charpy three-point bending test – Instron Ceast 9050.

The results of the tests are (i) magnitudes of the energy needed for breaking specimens in three-point bending loading mode and (ii) records of the force – time dependence upon impact mode in three-point bending. All results are presented in table 3 and figure 7. Peak forces achieved during the impact process were evaluated and summarized in table 3. During the pilot testing, seven samples were used per one group of the sandstone.

The results show clearly positive effect of consolidation by the strengthener REMMERS KSE 300 E from the viewpoint of sandstone dynamic mechanical properties. After consolidation of specimens the average energy, consumed for breaking a specimen, was increased by almost 20%. Considering the natural origin of the material tested the scatter of the results was not distinctive and the results obtained can be qualified as representative. From the viewpoint of force peaks attained at the tests it can be stated that due to the consolidation their average magnitude was increased by 62% as compared to non-consolidated material. However, the scatter of force peaks was somewhat greater. A higher scatter of resulting magnitudes is caused most probably by a lower geometric homogeneity of sandstone grains in the location of fracture which is characteristic for this type of natural materials [14] and [15]. As can be seen in figure 7 there is also some difference in the shape of "force-time" curves. After consolidation the dynamic E modulus of sandstone became higher and thus the material after consolidation became stiffer. The instrumented Charpy tests for this type of material appear to be functional, but however, until now no standard was defined which would clearly specify this procedure (specimen size, the way of performance, etc.). The testing method can also form an appropriate supplement to other dynamic test methods as it is shown e.g. in [16].

**Table 3.** Results – Charpy test.

| Specimen no. | Total energy [J] | Peak force [N] | Mass gain by treatment [w%] | Product gel deposition rate [w%] | Description | Comment |
|---|---|---|---|---|---|---|
| 1 | 0.109 | 729.7 | - | - | notched without cons. | |
| 2 | 0.112 | 1173.5 | - | - | notched without cons. | |
| 3 | 0.115 | 938.7 | - | - | notched without cons. | |
| 4 | 0.111 | 991.9 | - | - | notched without cons. | |
| 5 | 0.115 | 1037.4 | - | - | notched without cons. | |
| 6 | 0.111 | 631 | - | - | notched without cons. | bigger grain in frac. |
| 7 | 0.109 | 658 | - | - | notched without cons. | |
| Mean | 0.112 | 880.0 | - | - | | |
| Stand. deviation | 0.0023 | 193.0 | - | - | | |
| 8 | 0.135 | 1560.8 | 2.03 | 35 | notched consolidated | KSE 300E |
| 9 | 0.123 | 1016.8 | 1.73 | 32 | notched consolidated | KSE 300E |
| 10 | 0.128 | 1257.2 | 1.77 | 33 | notched consolidated | KSE 300E |
| 11 | 0.136 | 1535.2 | 1.91 | 34 | notched consolidated | KSE 300E |
| 12 | 0.14 | 1707.1 | 1.51 | 30 | notched consolidated | KSE 300E |
| 13 | 0.132 | 1147.3 | 1.44 | 30 | notched consolidated | KSE 300E |
| 14 | 0.145 | 1767.4 | 1.48 | 32 | notched consolidated | KSE 300E |
| Mean | 0.134 | 1427.4 | 1.696 | 32.286 | | |
| Stand. deviation | 0.0068 | 267.1 | 0.2107 | 1.7496 | | |

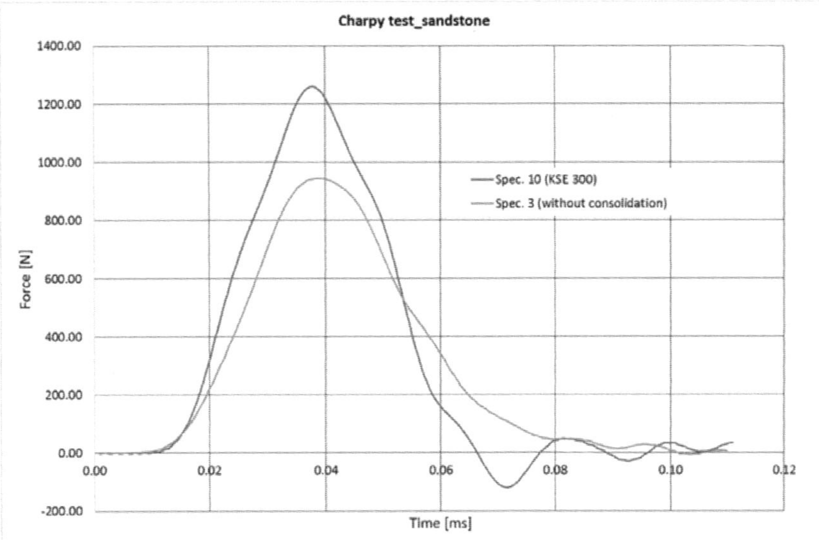


**Figure 7.** Examples of „Force – Time" dependences during Charpy test with or without consolidation.

Further experimental research on the effect of consolidation on dynamic resistance of sandstone material will be performed. Further tests of unnotched specimens as well as notched specimens with a different shape of the notch will be carried out as well.

## 4. Conclusions

Two procedures for impact testing of building materials were introduced, including the pilot tests results. The feasibility and usefulness of the procedures for characterization of the sandstone and concrete samples was assessed. The influence of consolidator REMMERS KSE 300 E on the impact resistance of the sandstone specimens was also evaluated. The following conclusions can be drawn:

- Results of the dynamic tensile strength tests showed feasibility and usefulness of the proposed test method for description of the mechanical properties of the cement-based materials under impact load. Tested specimens with lower tensile strength under static load showed lower characteristics under impact load as well. The method also showed consistent results in terms of coherency of achieved total fracture energy and peak force.
- Adjustments in the shape parameters of the test specimens are required due to a higher sensitivity of concrete to stress concentration in the notches. Specimens with a zero-gauge length were proposed for further testing, as they seem to be more suitable for designed test configuration.
- Regarding the Charpy test of sandstone, the test method was evaluated as suitable to assess the dynamic resistance of quasi-brittle inelastic silicate composite materials.
- The positive influence of consolidator REMMERS KSE 300 E on the impact resistance of the sandstone was proven; the average consumed energy, needed to break a specimen, was found to be increased by almost 20% for consolidated specimens. Also, due to the consolidation process the average force peak was increased to 1427.4 N, i.e. by 62%, as compared to non-consolidated material.

Although used testing methods are not standardized for the quasi-fragile materials, their application for a further characterization of materials can be recommended.

**Acknowledgments**

The authors acknowledge kind support from the Technological Agency of the Czech Republic (Project No. FW01010021).

**References**

[1] Thomas R J and Sorensen A D 2018 Charpy Impact Test Methods for Cementitious Composites: Review and Commentary *Journal of Testing and Evaluation* Vol. **46** No. 6 pp 2422–2430
[2] Šperl M and Drdácký M 2018 Non-standard experimental tests of sandstone and its pre-cracking for fracture testing *IOP Conf. Ser.: Mater. Sci. Eng* **379** 012029 pp 1–7
[3] Grote D L, Park S W and Zhou M 2001 Dynamic Behavior of Concrete at High Strain Rates and Pressures: I. Experimental Characterization *Int. J. Impact Eng*. Vol. **25** No. 9 pp 869–886
[4] Penelis G G 2019 *Concrete Buildings in Seismic Regions* (ed. Taylor &Francis group) p 775 ISBN 978-1-138-10687-1
[5] Keyes C D et al. 2005 *Medical Response to Terrorism Preparedness and Clinical Practise* (ed. Lippincott Williams & Wilkins) p 453 ISBN 0-7817-4986-7
[6] ISO 8256:2004: Plastics – Determination of Tensile-Impact Strength
[7] ASTM E23-16b:2016 Standard Test Method for Notched Bar Impact Testing of Metallic Materials (West Conshohocken: ASTM International)
[8] ISO 148-1:2009 Metallic Materials—Charpy Pendulum Impact Test (Geneva, Switzerland: International Organization for Standardization)
[9] ASTM D6110-10:2010 Standard Test Method for Determining the Charpy Impact Resistance of Notched Specimens of Plastics (West Conshohocken: ASTM International)
[10] ISO 179-1:2010 Plastics—Determination of Charpy Impact Properties Geneva, Switzerland: International Organization for Standardization)
[11] Drdácký M, Šperl M and Jandejsek I 2016 Consolidation effects on sandstone toughness *Proceedings of the 13th Int. Congress on the Deterioration and Conservation of Stone - SCIENCE and ART: A Future for Stone* Vol. **2** pp 687–694 (Paisley, UK: University of the West of Scotland)
[12] *Online* https://www.remmers.co.uk/en_GB/constructionflooring/conservation-restoration/kse-modula-system/c/c6796
[13] Suksawang N, Wtaife S and Alsabbagh A 2018 Evaluation of Elastic Modulus of Fiber-Reinforced Concrete *ACI Materials Journal* **115** (2) pp 239–249
[14] Hu S, Tan Y, Zhou H et al. 2017 Impact of Bedding Planes on Mechanical Properties of Sandstone *Rock Mech Rock Eng.* **50** pp 2243–2251
[15] Remzova M, Frankeova D, Slizkova Z, et al. 2016 Effect of modified ethylsilicate consolidants on the mechanical properties of sandstone *Construction and Building Materials* **112** pp 674–681
[16] Barr B and Bouamrata A 1988 Development of a Repeated Dropweight Impact Testing Apparatus for Studying Fibre Reinforced *Concrete Materials Composites* Vol. **19** No. 6 pp 453–466

# Chemical stability of innovative grouts for cast basalt elements

**T Zlebek[1], J Hodul[1], L Meszarosova[1] and R Drochytka[1]**

[1] Department of Building Materials and Components, Faculty of Civil Engineering, Brno University of Technology, Veveri 331/95, Brno 602 00, Czech Republic

Email: zlebek.t@fce.vutbr.cz

**Abstract**. Newly developed tile grouts are intended mainly for environment with strong chemical loading. For this reason, chemical resistance was one of the key parameters observed. The grout for cast basalt tile was polymer based, in particular with epoxide binder and it utilized solidified hazardous wastes as fillers. It is well-known that epoxide binder has high resistance to solutions of inorganic substances. However, it is easily degraded by organic compounds. This was taken into consideration when assessed aggressive environments and their concentrations were selected. Examined substances were 60% solution of sulfuric acid, vapors of this acid (at concentration 96%) and 10% solution of acetic acid. After exposition to action of these aggressive substances, first of all the visual changes were assessed, i.e. change of color, formation of blisters, cracks, flaking or change of sheen. Then influence on microstructure, changes of bulk density and strengths of materials were observed.

## 1. Introduction

In the building industry there are many possibilities, how to prepare tile grouts. Usually are grouts prepared with inorganic cement binder [1]. There is possibility to use polymers in the process in preparation of grouts. Epoxy resins are one of the widely used resins to produce polymer- modified concretes and mortars [2]. Epoxy grouts are used in applications with required high strength, rapid setting, handling versatility dynamic, load bearing, critical alignment and resistance against aggressive chemical environment [3, 4, 5]. It is well-known that epoxide binder has high resistance to solutions of inorganic substances. However, it is easily degraded by organic compounds. Tile grouts should have maximal adhesion to the base and high tensile and shear strength, rather than other mechanical properties [6]. For very good properties of epoxy resins, it is necessary to pay quite high price. To mitigate the high cost of epoxy resins, a range of fillers can be added to dilute the resin content [7]. For example, very fine graded fillers can be applied into polymer concrete – most of all milled stone materials (for example quartz dust, silica fume, granite meal) [8]. With growing importance of sustainable development requirements is necessary to use wastes and by-products to save raw materials. Fly ash is one of the commonly used filler in epoxy-based polymer concrete [8].

## 2. Materials and Methods

The input materials are fillers in the form solidification product (SP) and the epoxy resin as polymeric binder. The first one of fillers (SP1) consists of the quartz dust (50%), the fly ash form high-temperature combustion (40%) and the end product as the hazardous waste from the municipal solid waste incineration, the other one (SP2) consists of the quartz dust (50%), the fly ash form high-temperature combustion (40%) and the hazardous waste in form of cement bypass dust from the cement plant. These solidification products were incorporated into the epoxy grouts, and subsequently observed in chemically aggressive environment and compared with reference filler consisted from quartz dust. The standard EN 12808-1 [9] does not give exact definition of limit values for tile grouts chemical resistance, nor chemical substance to be used for resistance tests. However, substance to which the chemically resistant grouts are usually exposed, should be used as testing agent. Similarly, the test conditions (like temperature or humidity) should be close to design conditions, where tested grouts will be applied (sewers and industrial plants with chemical aggressive environment).

Selected chemically aggressive environments and test conditions to which the samples of tile grouts will be exposed:

- 96% sulfuric acid ($H_2SO_4$) p. a.: above surface of $H_2SO_4$ (in vapors), at temperature 60°C,
- 60% solution of $H_2SO_4$: submersion to solution of $H_2SO_4$ at 23°C,
- 10% solution of acetic acid ($CH_3COOH$): submersion to solution at 23°C.

## 3. Results and discussion

Samples made from the tile grout were measured and weighed before test of chemical resistance. Samples for exposition in vapors of $H_2SO_4$ were then placed on ceramic grid above surface of 96% $H_2SO_4$. Glass vessel with samples and chemical agent was the closed and placed for 14 days in drying room, where the temperature was kept constantly at 60°C. Samples for exposition to 60% solution of $H_2SO_4$ and in 10% solution of $CH_3COOH$, were placed on the bottom of desiccator figure 1. Individual chemical agents were added in the vessels figure 2; the vessels were closed and stored for 14 days at 23°C.

(a) (b)

**Figure 1.** Location above surface of $H_2SO_4$ (in vapors); samples placed in the glass desiccator (a), on ceramic grid above surface of 96% $H_2SO_4$ (b).

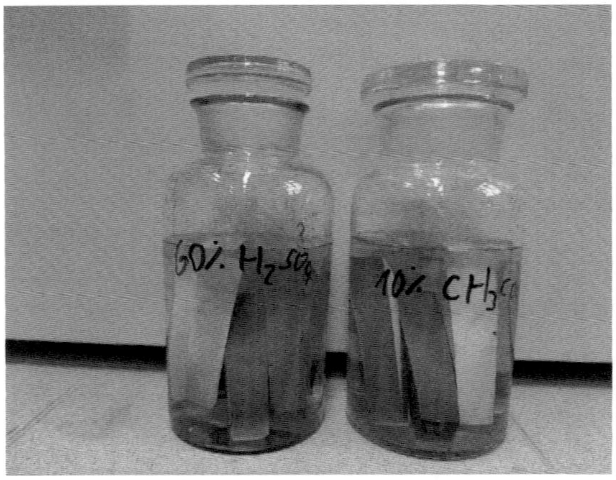


**Figure 2.** Testing samples placed in 60% solution of $H_2SO_4$ and 10% solution of $CH_3COOH$.

Selection of chemical agents and testing environments were maximally chosen close to conditions determined by environment, where the materials will be applied in practice – the grout utilization is intended for chemical aggressive environment with very low pH, up to value of 1, what represents very acid aggressive environment.

After selected time of exposition, the samples were taken out of the vessel with individual chemical substances, dried and visually assessed figure 3. It was found out that samples with all used types of fillers (SP1, SP2 and REF) showed no visual changes after exposition to 96% vapors of sulfuric acid at the temperature 60°C. Direct contact of solution of sulfuric acid with the sample caused color changes on the surface, which could be characterized as lighter color shade. Considerable color changes were observed on the samples containing the fillers SP1 and SP2 exposed to solution of acetic acid. Influence of this acid made the color shade lighter and many stains were formed, probably by the precipitation of salts on the surface. Less marked color and visual degradation was observed on the sample containing the filler SP1, which seems to be more suitable from the chemical resistance point of view.

(a) (b) (c)

**Figure 3.** Samples after the chemical resistance test: left side – reference samples, in the middle – samples with the filler SP2, right side – samples with the filler SP1; (a) samples exposed to the 10% acetic acid solution, (b) samples exposed to 60% solution of sulfuric acid, (c) samples exposed to the vapors of the 96% sulfuric acid.

Samples of the tile grout exposed to chemically aggressive environment were subjected to the microstructure observation by the digital optical microscope. Figure 4 shows clearly progress of color change on the surface of the sample; lighter color shade propagated from the surface to the core of the sample, while darker core is visible in original color. Detailed view of degraded surface layer is visible in figure 5. Color change propagating from the edge of the sample to the middle is not so significant on samples exposed to acetic acid (figure 6). Degradation is observed locally, in separated sections, formed probably by precipitated acetates, which is apparent in the detail of degraded surface in figure 7. It was stated by Kalenda and Kalendova [10] that $H_2SO_4$ is strong acid, so it is almost completely dissociated, and its ions are not able to enter the hydrophobic resin so that attack by this acid takes place only in the surface layer. The organic acids, such as $CH_3COOH$ and $CH_3CH_2COOH$ are relatively weak acids and much less dissociated – they act as solvents and their reaction with epoxy resin resulting in swelling, and segments of the macromolecularresin chains are separated.

**Figure 4.** Sample containing the filler SP2 after submersion in the solution of 60% $H_2SO_4$ — cross section.

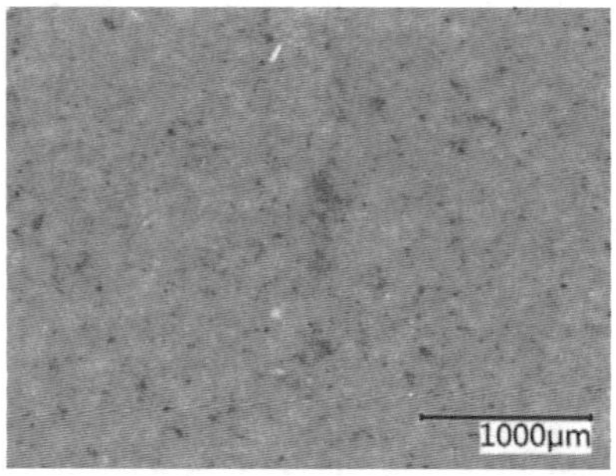


**Figure 5.** Sample containing the filler SP2 after submersion in the solution of 60% $H_2SO_4$ — surface.

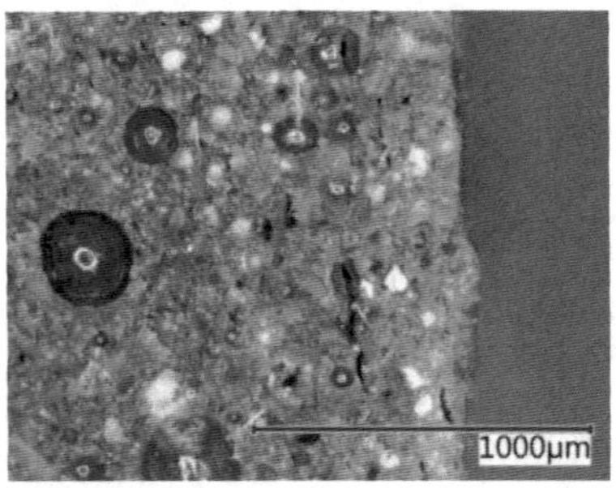


**Figure 6.** Sample with the filler SP2 after submersion in 10% $CH_3COOH$ — cross section.

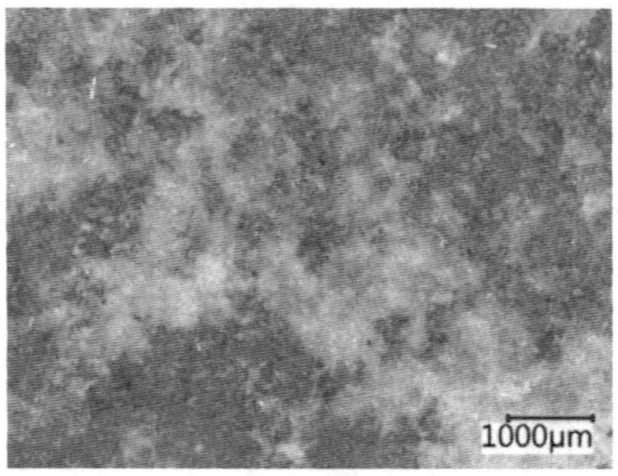


**Figure 7.** Sample with the filler SP2 after submersion in 10% $CH_3COOH$ — surface.

Subsequently, the grout samples were measured and weighed to determine bulk density and for comparison of samples exposed to aggressive environment to those placed only in laboratory conditions. The results of bulk density are summarized in table 1 and figure 8.

**Table 1.** Influence of the exposition to aggressive environment to the bulk density of samples.

| Bulk density [kg/m³] | | | | | | | | | | | |
|---|---|---|---|---|---|---|---|---|---|---|---|
| Laboratory conditions | | | After exposition to 96% solution of sulfuric acid vapors at 60°C | | | After submersion in the solution of 60% $H_2SO_4$ | | | After submersion in the solution of 10% $CH_3COOH$ | | |
| SP1 | SP2 | REF | SP1 | SP2 | REF | SP1 | SP2 | REF | SP1 | SP2 | REF |
| 1370 | 1370 | 1260 | 1380 | 1370 | 1250 | 1390 | 1390 | 1250 | 1360 | 1360 | 1240 |

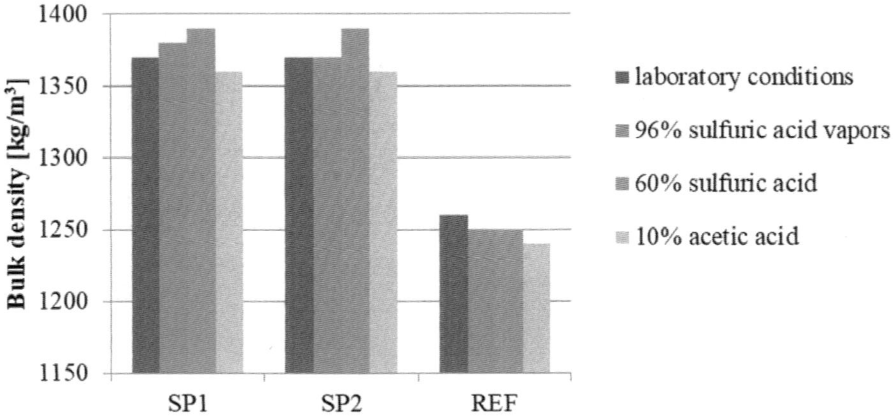


**Figure 8.** Influence of the exposition to aggressive environment to bulk density change.

The results show the trend of the bulk density reduction of samples with reference filler, caused by action of chemically aggressive environment. Other fillers showed opposite influence and exposure to the sulfuric acid in direct contact with samples caused increase of bulk density; by 10% in the sample containing SP1 filler. Influence of vapors of this acid caused no changes of bulk density. Action of acetic acid caused reduction of bulk density of all tested materials by approximately 10 to 20 kg/m$^3$. Hence, it can be stated that selected aggressive environments have no significant influence on change of bulk density.

These samples were then subjected to tests of three-point flexural strength and fragments of the samples were used for the determination of compressive strength. Both tests were performed in accordance with EN 12808-3 European harmonized standard [11]. The results of strengths are summarized in tables 2–3 and in figures 9–10. Properties after exposition to aggressive environment were compared to the reference samples, placed only in laboratory conditions.

**Table 2.** Influence of the exposition to aggressive environment to flexural strength.

| Flexural strength [N/mm$^2$] | | | | | | | | | | | |
|---|---|---|---|---|---|---|---|---|---|---|---|
| Laboratory conditions | | | After exposition to 96% solution of sulfuric acid vapors at 60°C | | | After submersion in the solution of 60% $H_2SO_4$ | | | After submersion in the solution of 10% $CH_3COOH$ | | |
| SP1 | SP2 | REF | SP1 | SP2 | REF | SP1 | SP2 | REF | SP1 | SP2 | REF |
| 33.8 | 36.4 | 42.3 | 33.5 | 39.0 | 37.6 | 35.0 | 38.5 | 44.8 | 34.2 | 33.2 | 38.5 |

**Table 3.** Influence of the exposition to aggressive environment to compressive strength.

| Compressive strength [N/mm$^2$] | | | | | | | | | | | |
|---|---|---|---|---|---|---|---|---|---|---|---|
| Laboratory conditions | | | After exposition to 96% solution of sulfuric acid vapors at 60°C | | | After submersion in the solution of 60% $H_2SO_4$ | | | After submersion in the solution of 10% $CH_3COOH$ | | |
| SP1 | SP2 | REF | SP1 | SP2 | REF | SP1 | SP2 | REF | SP1 | SP2 | REF |
| 54.4 | 51.5 | 44.2 | 60.0 | 47.5 | 50.0 | 60.0 | 55.0 | 54.0 | 56.3 | 47.5 | 48.5 |

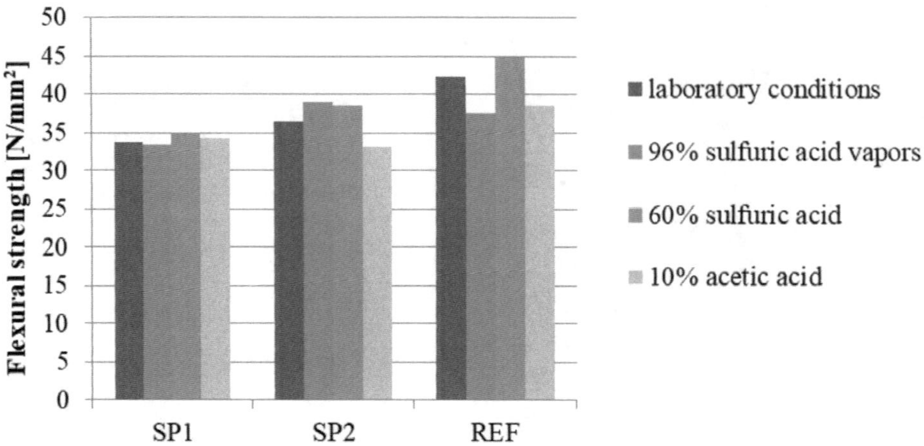


**Figure 9.** Influence of the exposition to aggressive environment to flexural strength of grout samples.

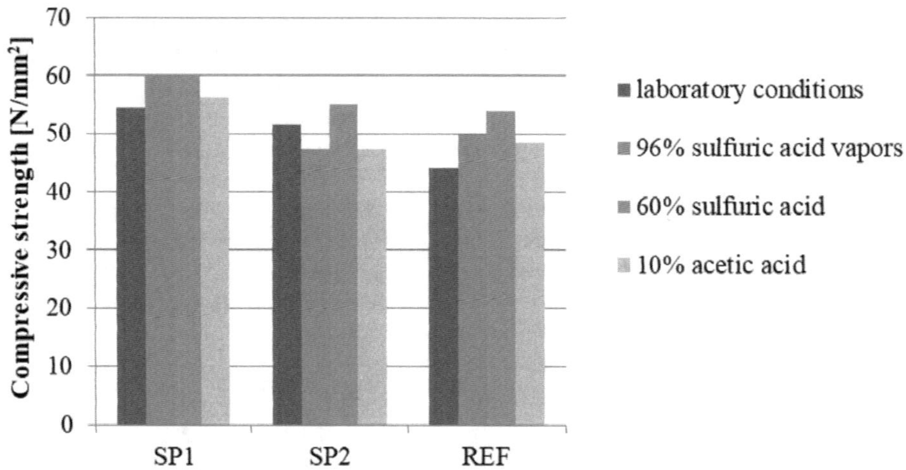


**Figure 10.** Influence of the exposition to aggressive environment to compressive strength of grout samples.

## 4. Conclusion

The influence of chemical aggressive environment on microstructure, changes of bulk density and strengths of polymeric grout materials were observed and evaluated. The results showed the trend of reduction of bulk density of samples with reference filler, caused by exposure to the chemically aggressive environment. The grouts containing solidification product as the filler showed similar compressive and flexural strength with the reference sample containing only quartz dust as the filler. It is not possible to clearly state the negative influence of exposition of individual samples to various chemical environment based on the results of compressive and flexural strength. The grout with the filler SP1 (the mixture of quartz dust, fly ash and end product) showed quite better strengths than the reference samples. Specimens showed visual changes after exposition to sulfuric acid (60% solution) and acetic acid (10% solution). It can be stated that the use and implementation of specifically prepared fillers (solidification products containing hazardous waste), showed the physical and mechanical properties comparable to the reference material. It could be possible to use such material in defined conditions with aggressive environment, in this case with attack of very low pH, up to the value of 1 (60% $H_2SO_4$).

**Acknowledgements**
This paper was supported by the project FAST-S-20-6136 "Study of the incorporation of hazardous waste and secondary raw materials in a polymer matrix of special composite materials".

**References**

[1] Palardy D, Onofrei M and Ballivy G 1998 Microstructural changes due to elevated temperature in cement based grouts *Advanced Cement Based Materials* **8(3–4)** pp 132–138
[2] Anagnostopoulos C A, Sapidis G and Papastergiadis E 2016 Fundamental properties of epoxy resin-modified cement grouts *Construction and Building Materials* **125** pp 184–195
[3] Shamsuddoha M, Islam M M, Aravinthan T, Manalo A and Lau K 2013 Characterisation of mechanical and thermal properties of epoxy grouts for composite repair of steel pipelines *Materials & Design* **52** pp 315–327
[4] Kneuer R L and Meyers M 1991 Strengths and limitations of epoxy grouts *Concrete international* **54** p 6
[5] Prolongo S G, del Rosario G and Ureña A 2006 Comparative study on the adhesive properties of different epoxy resins *Int J Adhes Adhes* **26** pp 125–32
[6] Luso E and Lourenço P B 2017 Bond strength characterization of commercially available grouts for masonry *Construction and Building Materials* **144** pp 317–326
[7] Ferdous W, Manalo A, Wong H S, Abousnina R, AlAjarmeh O S, Zhuge Y and Schubel P 2020 Optimal design for epoxy polymer concrete based on mechanical properties and durability aspects *Construction and Building Materials* **232** pp 117–229
[8] Garbacz A and Sokołowska J J 2013 Concrete-like polymer composites with fly ashes – Comparative study. *Construction and Building Materials* **38** pp 689–699
[9] EN 12808-1:2008 Grouts for tiles. Part 1: Determination of chemical resistance of reaction resin mortars *European Committee for Standardization (CEN)* (Brussels, Belgium)
[10] Kalenda P and Kalendova A 1994 Improved chemical resistance of epoxy resin-based coating compositions *Dyes and Pigments* **27(4)** pp 305–312
[11] EN 12808-3:2008 Grouts for tiles - Part 3: Determination of flexural and compressive strength *European Committee for Standardization (CEN)* (Brussels, Belgium)

# AUTHOR INDEX